SUPERCHARGE YOUR BRILLIANCE!

Supercharge Your Brilliance!

The Modern Businesswoman's Guide to Scaling Faster with AI

TAMIKO L. CUELLAR

PYP
BOOKS

PYP Books

CONTENTS

First published in 2023 by
PYP Books, a division of Pursue Your Purpose, LLC
St. Louis, Missouri
www.PursueYourPurpose.com

ISBN-13: 979-8-9878795-0-4 (paperback)
ISBN-13: 979-8-9878795-1-1 (e-book)

Cover and interior design by PYP Books
St. Louis, Missouri USA

First edition: 2023

Printed in the United States of America.

Preface

Hey there! With great joy and excitement, I present to you my labor of love, *Supercharge Your Brilliance! The Modern Businesswoman's Guide to Scaling Faster with AI*. I wrote this book because of my deep passion for empowering women entrepreneurs. You deserve to work smarter and not harder in your business and use every advantage to succeed.

I understand what it feels like to start a business with limited resources and a lack of mentorship. When I embarked on the journey to launch my own company over a decade ago, I faced seemingly insurmountable challenges but persevered and succeeded. Now, 12+ years later, I've integrated artificial intelligence (AI) into many of my business practices, which has been a game-changer! But, I made a glaring observation as I began searching for entrepreneurs using AI in their companies. There were not a substantial number of women in the AI space!

Even as I watched countless YouTube videos and searched online for books and courses, it became clear that men dominate the AI space. The tech developers, AI companies, and self-proclaimed gurus were predominately men. This observation confirmed that I needed to jump into the AI space to help businesswomen avoid getting left behind while ensuring they'd become early adopters of this incredible technology. AI is about to become more mainstream, and the business world is on the cusp of using it to change their customers' lives and how companies do business.

So now, I am paying it forward with my 5th book by helping other women in business shorten the learning curve with some fantastic business growth hacks. I want all entrepreneurs, especially my fellow businesswomen, to win, and I know that AI will play a significant role in making that happen. I'm convinced that AI will help level the playing field by allowing women entrepreneurs to scale their organizations faster. This book is my way of sharing this message with the world.

You see, I'm not new to embracing technology. I genuinely love technology and its ability to make our lives easier and scale our businesses faster. In fact, in 2013, I wrote and published a mini guide on Amazon titled "101 Tools to Take Your Startup or Solopreneur Business to the Next Level." It highlighted technology tools that can help entrepreneurs to grow. Then, in 2015 I partnered with Microsoft to launch a workshop series for entrepreneurs called "Leveraging Technology to Grow Your Business." This workshop series showcased the power of technology tools and how to use them to achieve business growth. I scaled my company to four other countries and recruited, hired, and managed a team of individuals from five countries primarily because of the capabilities of technology. So, yes, I am a believer!

I am a proud Gen-X'er, so I grew up watching my parents play eight tracks and vinyl records and use rotary phones. I still remember using floppy discs, cassette tapes, VCRs, push-button phones, typing green letters on a black computer screen as a kid, and waiting an eternity to download a simple image using dial-up. In college, my first cell phone was a car phone because it could only be used in the car and had a battery pack the size of a Bible! Ahhh, the good ol' days! So, just like I've ridden many waves of technological advancements, I'm riding the wave of AI into the future.

AI is the transformation we never knew we needed for our businesses. It's revolutionary! AI "supercharges" your brilliance by augmenting your existing capabilities by helping you do essential tasks 1,000 times better and faster. In this book, I share over 100

AI tools you can use in your business to automate tasks, streamline processes, save you money, make you money, and scale fast.

I sincerely hope and pray that this book will reach as many women entrepreneurs worldwide who have yet to discover the power of AI and elevate their organizations to the next level. A few simple AI tools might be the keys to unlocking your next level! So, thank you for investing in yourself. I know that by doing so, you have made a powerful step toward achieving your business goals, and I am honored to be a part of your journey. Your brilliance is about to be supercharged!

Introduction: What's this Book All About?

Are you ready to level up your business game and maximize your company's efforts with half the work and half the money? Then buckle up because we're about to explore what it means to supercharge your brilliance as a modern businesswoman!

First things first, what is artificial intelligence? Artificial intelligence, or AI, is the simulation of human intelligence in machines designed to think and act like humans. AI is a field of computer science that focuses on creating devices and software that can perform tasks that typically require human intelligence. These include recognizing images, understanding natural language, making decisions, and solving problems.

AI aims to build machines that can think and act like humans but with as little need for human intervention as possible. AI systems use algorithms, which are sets of instructions given in a programming language, to process and analyze data and make predictions or better choices based on that information. These algorithms are trained on large amounts of data to improve their accuracy over time.

Examples of popular AI applications are Siri, Alexa, and Google Assistant, which are virtual assistants that can answer questions and perform tasks using natural language processing and other AI technologies. Other examples include self-driving cars, which use computer vision and machine learning to navigate roads and make driving decisions, and recommendation systems, which use AI to suggest products or services to users based on their preferences. And don't worry about understanding the AI tech jargon! You don't need to know these terms to use AI effectively to scale your company or enhance your everyday life. You may refer to the glossary in the Appendix to further your knowledge of these terms.

Now let's talk about what we mean by "brilliance." It isn't just about being smart or good at running your business. It's about taking unique skills, qualities, and experiences that make you stand out in your field and multiplying it times 1,000! It's about augmenting your strengths, being more confident in your abilities, and constantly working to improve and grow.

So, how do you take your brilliance to the next level? That's where the "supercharging" part comes in. Think of it like turbocharging a car engine - adding extra power to get ahead. To supercharge your brilliance as a modern businesswoman, you must embrace new technologies and innovative approaches to help you reach your goals faster and more effectively.

Enter AI. Artificial intelligence has revolutionized how the world conducts business, and it's time for you to take advantage of it. From automating repetitive tasks to analyzing information and making predictions, AI can help you streamline your operations, achieve better outcomes, save time and money, and scale your organization much faster.

Who will benefit from this book?

Supercharge Your Brilliance! is the guide for modern, unconventional businesswomen who want to scale their businesses smarter and faster with new AI tools that cut the work of human labor in half and, in many ways, make the work that we humans do even better. It's perfect for women who run their own businesses, whether it's a solo gig or a big company. I've even carved out a special little section for nonprofit organizations. This book covers everything you need to jumpstart your understanding of AI and how it can help you rapidly grow your organization. From streamlining operations to boosting profits, you'll learn practical ways to harness this emerging technology to improve your workflow, marketing strategies, and more.

I also wrote this book to empower and equip businesswomen in leadership roles inside the workplace with tools they can implement to do their jobs better. If you manage a team, you can introduce them to AI tools that will make them more efficient and make you and your team look like rock stars! The information in this book can inspire change within your workplace. With a better understanding of AI and its capabilities, you'll be able to pitch new and innovative ideas to upper management to improve the overall workflow of your organization. If you're unsure how to do so, give your boss a copy of this book! You'll benefit yourself, your team, and the entire company while positioning yourself as a change agent. This guide empowers you to become a leader in your field and drive digital transformation in your workplace.

Suppose you are an independent consultant, gig worker, graphic artist, copywriter, marketing guru, editor, social media influencer, YouTuber, or another type of creative. In that case, you will also find something to apply to your work to make you much more productive. Not only that, but if you're a creative professional, this guide will also give you an edge over your competition. You'll learn to use AI to replace some human tasks and free up more time to attend to what you do best - creating. You'll also discover new and innovative ways to market your work and reach a wider audience. The practical information in this guide will help you streamline your workflow, increase efficiency, and increase your target market.

Are you a technology enthusiast who has an affinity for all things tech? Then you will find aspects of this book valuable and enjoyable also. Not only is this book perfect for women business owners, businesswomen in the workplace, and creatives, but tech enthusiasts will also find it engaging and informative. This guide provides an in-depth look at AI and its applications in the modern world, making it a must-read for anyone passionate about technology. You'll learn about the latest advancements in AI and how they are being used to revolutionize various industries. This book is an easy-to-understand guide, so you don't have to have a background in technology to benefit from it. Whether you're a student, hobbyist, or professional in the tech field who is just passionate about technology, this guide will give you more insight into AI and its potential.

What can you expect to get out of reading this book?

You'll discover how to take the intimidation out of data analysis and customer relationship management for your organization by using AI tools to automate more duties, freeing up resources to devote to your core business. You'll also learn how to effectively reach more of your target market and grow your brand with less work.

You'll find real-world examples and AI platforms you can use, so whether you're tech-savvy or just starting, you'll find it easy to follow. It's not just about tech, however. It's a call to action for women entrepreneurs to embrace their brilliance and use new technology to reach their goals. You'll learn AI digital marketing tools to reach a wider audience and stay ahead of the competition. It has all the necessary information and advice to jumpstart your next level.

Moreover, this guide will help you better understand AI and provide case studies of how companies use AI today. You can take these tools, research them, put them to the test, and see the results of using AI in your business almost immediately. Whether you're new to AI or have been using it for a while, this guide will provide valuable insights and information to help you scale your business faster and further. So, if you want to gain a competitive edge, increase efficiency, and grow your business, *Supercharge Your Brilliance!* will be an excellent resource for you. Get ready to unlock your potential and become an AI-powered business leader.

Do Women Entrepreneurs Need to Use Technology More?

Sadly, women entrepreneurs are often less likely to use technology than their male counterparts. Books like this are still necessary as a call to action for all serious women entrepreneurs to take their brilliance and businesses to the next level with AI. The reasons that women-owned firms are lagging behind their male counterparts in using AI include limited access to tech, limited technical skills, cultural and societal biases, and more. Studies have shown that women-owned firms are falling behind in using tech and AI compared to male-owned firms. Overcoming these barriers and empowering women to embrace technology and AI will be critical to helping them succeed in the 21st century.

- A 2019 survey by Sage found that 32% of women small business owners had no plans to use AI, compared to only 23% of male small business owners (Sage, 2019).
- According to a 2019 National Center for Women & Information Technology (NCWIT) study, women-owned small businesses are less likely to use AI for customer insights and marketing automation than male-owned small businesses. The study found that only 11% of women-owned small companies use AI for these purposes, compared to 18% of male-owned small businesses (NCWIT, 2019).
- A 2021 National Women's Business Council (NWBC) report found that women-owned businesses are less likely to use

technology to expand their businesses. Only 23% of women-owned enterprises use digital platforms for e-commerce, compared to 37% of all small businesses (NWBC, 2021).

- Men tend to use artificial intelligence more frequently than women. The AI Index report indicates that 71% of AI users are male, highlighting a gender gap in technology-related fields not only in terms of employment but also usage (M.,2023).

- In 2022, the World Economic Forum (WEF) reported that only 2% of venture capital funding in 2019 was allocated to AI startups founded by women. Women are significantly underrepresented in the field of AI, accounting for only 22% of AI professionals globally, 13.83% of AI paper authors, and 18% of authors at leading AI conferences (WEF, 2022).

- Disruptive technologies can empower women to overcome barriers of economic participation by providing new work opportunities and enabling women entrepreneurs to compete on an equal footing with men. However, this is only possible if women have access to technology. The digital divide is a problem globally, with men being 20% more likely to be on-line and 50% more likely in less developed countries. There-fore, it is critical that women have equal access to technology to thrive as leaders, founders, and employees in the digital economy (Rubin, Hakspiel & Gray, 2021).

Overall, while there have been some improvements in women's adoption of technology in their companies, there is still a signifi-cant gender gap regarding general technology use. Addressing this disparity by empowering women entrepreneurs to embrace tech-nology and AI will be essential to helping them succeed and grow their organizations. As the business world continues to evolve, technology has become an increasingly important tool for every entrepreneur.

For women entrepreneurs, in particular, technology can provide higher efficiency, reach, and the impact that can help them overcome traditional barriers to success and achieve their goals. Some traditional barriers include gender bias, lack of access to funding, limited access to networks, and difficulty balancing work and family responsibilities. While AI is not a silver bullet to solve all of these challenges, it can definitely help to address some of them. Here are some potential ways AI can help:

Mitigating gender bias in funding opportunities. AI can be used to screen pitch decks and other funding applications and identify qualified candidates based on their skills, experience, and creditworthiness rather than relying on potentially selective human judgment. Removing selective judgment can help increase the number of women selected for funding.

Enhancing networking opportunities. AI can help women business owners to connect with other entrepreneurs and potential business partners by analyzing their interests and backgrounds and making recommendations for networking events or online communities that are most relevant to them.

Automating time-consuming tasks. By automating tasks such as data entry, scheduling, and other jobs traditionally done by women, women entrepreneurs can free up more time to concentrate on strategic planning and business development. AI can help to alleviate some of the workload associated with running a company and support women business owners to achieve a better work-life balance, which statistically is high on the priority list.

The use of AI helps to level the playing field for women entrepreneurs because it is an avenue that provides better reach to a broader audience. With the rise of digital marketing channels, women entrepreneurs have easier access to a global audience and can connect with potential customers worldwide. AI can accelerate the expansion of their target audience and reach new markets, which is particularly important for women entrepreneurs who may face barriers to entry in more traditional business settings. In addition, technology can help women entrepreneurs to maintain a strategic advantage in their industries. With access to cutting-edge tools and platforms, women entrepreneurs can analyze information, monitor market trends, and make informed choices to drive their organizations forward. Staying on the cutting edge allows women entrepreneurs to stay ahead of the curve in a rapidly changing business landscape.

The importance of using technology for women entrepreneurs cannot be overstated. Technology, specifically AI, is crucial in helping women entrepreneurs overcome barriers in the business world. Whether it's finding available funding opportunities, uncovering an innovative product or service to take to market, expanding globally, or just finding creative ways to stay ahead of the competition, technology and AI are powerful tools that keep women on the cutting edge of their industries as significant players rather than being more marginalized as a result of not staying ahead of the technology curve. With the right tools and approach, women entrepreneurs can harness the power of technology to unleash their full potential and achieve their goals.

| 1 |

AI - The Next Big Thing
Is Here!

> "For more than 250 years the fundamental drivers of economic growth have been technological innovations. The most important of these are what economists call general-purpose technologies – a category that includes the steam engine, electricity, and the internal combustion engine. The most important general-purpose technology of our era is artificial intelligence, particularly machine learning." – **Erik Brynjolfsson and Andrew McAfee, authors of "Machine, Platform, Crowd: Harnessing our Digital Future"**

Get ready to be amazed! The next big thing has arrived, and it's called AI! That's right, it's time to say goodbye to tedious old business practices and hello to taking a quantum leap to scale lightning-fast. And when it comes to business, AI is like a secret weapon, or as I like to say, a secret superpower that can help you supercharge your brilliance and achieve exceptional results.

Imagine being able to craft the most engaging sales copy in a few minutes that attracts hundreds of new customers, answer your customers' simple questions at warp speed without wait times, or analyze mountains of historical customer data to make accurate product recommendations in a snap – all with little no human intervention or money. Well, now you can accomplish all of this with AI! And that's just the tip of the iceberg when it comes to what AI can do for your company. So, buckle up and get ready for a wild ride because the future of business is here, and it's powered by AI!

What is AI, and why is it like having a secret business superpower?

Artificial intelligence, commonly known as AI, is the latest buzzword in the business world, and for a good reason. It's like having a supercomputer or a team of super-intelligent robots working for you, helping you computerize job responsibilities and analyze information faster than you ever could. It can be described as a secret business superpower that can revolutionize how you run your business. Imagine freeing up your time from monotonous tasks and being able to direct attention to the strategic, creative, and innovative parts of your organization. That's the power of AI!

But what exactly is AI? In simple terms, AI is a computer system designed to perform tasks that typically require human intelligence. It can perform tasks such as recognizing patterns, making predictions, and solving problems. AI can analyze massive amounts of data, find solutions in real-time, and improve its performance over time. It's like having a super-smart assistant who never makes mistakes and never takes a day off!

You might be thinking, "But AI sounds like science fiction, and how can I even start using it in my business?" Don't worry. You don't need to be a tech genius to start using AI. There are plenty of AI tools and solutions available that are designed specifically for small businesses. From chat agents that can handle client service inquiries to predictive analytics that can lead to better problem-solving, there's an AI solution out there for just about every business need.

So, why is AI like having a secret business superpower? For starters, it can help you save lots of time and money. By automating tedious tasks, you can prioritize the more important aspects of your company, like increasing your customer base and expanding your product offerings. AI can also help you make better, data-driven solutions, leading to improved profits and increased success for your organization. AI is the ultimate business partner for doing business the smart way. It helps human brains make better decisions by streamlining workflows and boosting efficiency. Whether you're in marketing, finance, or operations, AI is taking the business world by storm and becoming a must-have for accelerated success. Think of it as a trusted business advisor who gives you the upper hand by making you even more brilliant at what you do. So, AI is more than just a buzzword. It's a real-life game-changer for any organization. By embracing AI, you can streamline your operations and achieve your company's goals faster than ever.

A brief history of AI and how it's evolved into what it is today

Let's take a quick trip down memory lane and step into a time machine to journey back to the days of Ancient Greece, where the idea of AI was born. The concept of AI has been around for thousands of years. The ancient philosopher, Aristotle, believed in automatons or self-operating machines. Who knew that his idea would one day lead to the creation of AI?

Fast forward to the 1950s, and AI, as we know it today, started to take shape. A group of researchers gathered at Dartmouth College to discuss the idea of creating a machine that could "think" like a human. This event is considered the birthplace of AI as a scientific field. From there, AI has evolved at an astonishing pace, leading to the creation of groundbreaking technologies, such as the first AI language processing system, ELIZA, and the first chess-playing computer program, Deep Blue.

But it wasn't until the late 1990s and early 2000s that AI truly exploded. Remember IBM's Watson AI system introduced to the world on the Jeopardy game show in 2011? It was the first time that AI could use algorithms to answer questions in natural language which triumphantly crushed its human competition. The wide-spread availability of information, advances in computer processing power, and the rise of the internet all contributed to the popularity of AI. Companies like Google and Amazon began incorporating AI into their businesses, using it for search algorithms and personalizing customer shopping experiences.

Today, AI has evolved into a multibillion-dollar industry, with applications in almost every sector, from healthcare to finance to retail. From virtual assistants like Siri and Alexa to self-driving cars, AI is becoming a ubiquitous part of our daily lives. The field of AI is constantly evolving, with new advancements being made all the time. The future of AI looks even brighter with the development of cutting-edge technologies like deep learning, machine vision, and natural language processing.

So, there you have it, a brief history of AI and how it's evolved into the powerful technology it is today. It's incredible to think about how far AI has come in just a few short decades and even more exciting to imagine what the future holds. Who knows? In a few years, we may have robots that can do our laundry and take our dogs out for walks! But for now, let's enjoy the high-tech advancements in AI that are helping us make our lives easier and more efficient every day, both in our personal lives and our businesses.

How AI works today

To explain how AI works for us today, I'll first have to break it down into two main parts: machine learning and deep learning.

Machine learning is like teaching a baby how to do things. When babies are first born, they don't know how to do anything. They need someone to teach them how to walk, talk, and do other things. Similarly, a computer doesn't know how to do anything when it is first turned on. It needs to be taught how to do things. In machine learning, the computer is given a task to do and a set of examples of how to do it. These examples are called "training data." The computer uses this training data to learn how to do the task. It does this by creating a mathematical model that represents the task. This model is then used to make predictions about new, unseen information.

To illustrate, let's say you want to teach a computer to recognize dogs in pictures. You would give the computer a set of images of dogs and a label that says "dog." The computer would then use this training data to learn what a dog looks like. It would do this by creating a mathematical model that represents the patterns in the pictures of dogs. Once the computer has learned what a dog looks like, you can give it new, unseen images and ask it to tell you whether or not there is a dog in the picture. The computer would then use its mathematical model to make a prediction about the new image. If the prediction is correct, the computer has learned to recognize dogs in pictures.

Deep learning is like machine learning on steroids. It is a type of machine learning that uses neural networks to model complex patterns in information. A *neural network* is a mathematical model inspired by the structure of the human brain. Just like the human brain is made of neurons connected to each other, a neural network similarly comprises nodes that connect. These nodes represent mathematical functions that are used to make predictions.

In deep learning, a neural network is trained using an extensive amount of training data. The training process involves adjusting the connections between the nodes in the network until the network can accurately make predictions about the training data. Once the network has been trained, it can make predictions about new, unseen data.

AI is powerful because it allows computers to do things that were once thought to be impossible. With AI, computers can learn from data and make predictions about new, unseen data. This ability to learn from data and make predictions is what makes AI so powerful. Another reason AI is so powerful is that it can automate tasks customarily done by humans. For instance, AI can automate duties like image and speech recognition, natural language processing, and problem-solving. This automation saves time and effort, reduces the risk of errors, and increases accuracy.

AI can also sort through lots of information much faster than humans, making it possible for AI to make predictions and choose the best course of action in real-time. I realize this might all sound like a sci-fi movie, but this advanced discovery of how to train computers and robots to think and behave as humans is both fascinating and somewhat otherworldly.

Keeping pace with technological advancements

In today's rapidly evolving world, technology plays a more critical role than ever in shaping how we live, work, and communicate.

The speed at which technology is advancing is dizzying, and it can be hard to keep up. But as businesswomen and professionals, we must stay on top of the latest technological advancements if we want to succeed.

Let's look at how technology has evolved over the years to understand the importance of keeping pace with technology. From analog to digital, from cable to streaming, from VHS cassette tapes to streaming movies on TV and handheld devices, from standalone physical alarm clocks to alarm clock apps on cell phones, from landlines to VOIP and Wi-Fi calling, and the list goes on and on. How we consume information and interact with each other has changed dramatically, and it continues to change at an astonishing rate.

One of the best examples of how technology has evolved is in the music industry. From the days of vinyl records to the current era of streaming services, the ways we listen to music have drastically changed over time. The introduction of the MP3 format revolutionized the music industry by making it possible to store and share extensive collections of music files electronically. And now, with the popularity of streaming services like Spotify and Apple Music, it's easier than ever to access millions of songs from anywhere in the world.

Another example of how technology has evolved is the way we communicate. Gone are the days of having to be at home to make a call. With the advent of cell phones and VOIP, we can communicate from anywhere, at any time. We can call, text, and video call from anywhere in the world, as long as we have an internet connection.

So why is it so important to keep pace with technological advancements? The simple answer is that we risk being left behind if we don't. In today's fast-paced world, organizations that fail to evolve with technology risk becoming obsolete. Just look at companies like Kodak, Blockbuster, and Blackberry, who were once leaders in their respective industries but failed to keep pace with technological advancements and were eventually left behind.

Here are ten examples of companies that became obsolete because they did not keep pace with technology:

1. Myspace was one of the first social networking sites, but it couldn't compete with the rise of Facebook.
2. Blockbuster was a well-known video rental chain that was replaced by the rise of online streaming services like Netflix.
3. Borders was a large bookstore chain that lost its competitive edge to e-books and online retailers like Amazon.
4. The Palm Pilot was once a popular personal digital assistant, but Palm was overtaken by the rise of smartphones.
5. AOL was once one of the largest internet service providers but lost its innovative edge in the marketplace to broadband internet.
6. Compaq - Compaq was once a leading computer manufacturer, but with the rise of laptops and tablets, Compaq needed help to keep up and eventually merged with Hewlett-Packard before its eventual demise.
7. Polaroid - Polaroid was once a leader in instant photography. Still, it was no match for digital cameras and has struggled to rebrand and make a comeback since its first bankruptcy in 2001.
8. Circuit City was once the largest electronics retailer in the United States, but with the rise of online electronics stores, it went out of business.
9. Toys R Us was the leading chain of toy stores in the U.S. Its brick-and-mortar retail stores could not keep pace with the accessibility of online toy retailers and was forced to close all of its 800 stores for good.

10. TomTom was once a popular brand of GPS navigation systems, but it couldn't compete with the rise of smartphones with GPS capabilities.

Staying on the cutting edge of technology advances such as AI is essential to remain relevant. Many colleges and universities globally are taking AI more seriously because they see its increasing impact on society, businesses, and the day-to-day lives of people in general. As such, universities like Northwestern University, DePaul University, and others now have advanced degree programs in AI. This is a good indicator that assures us that AI is here to stay for a very long time, or at least until the next generation of technology takes its place.

| 2 |

What Can AI Do for Your Business?

> "The playing field is poised to become a lot more competitive, and businesses that don't deploy AI and data to help them innovate in everything they do will be at a disadvantage." — **Paul Daughtery, Chief Technology & Innovation Officer of Accenture**

It's time to get up close and personal with AI! Here is where you'll get to know your "new best friend" and learn all about what it can do for your business. Think of AI as a powerful tool that can help you do more, and achieve more, but only faster. But before diving into the AI world, it's essential to understand exactly what AI is and how it works.

You will learn about AI's different uses and applications, how they can help your business, and their benefits in your day-to-day operations. You'll also learn about the available AI tools and resources and how to use them to supercharge your brilliance.

So, grab your thinking cap and get ready to learn because this chapter is your chance to get to know AI and see just what it can do for your business. Whether you're new to the game or a seasoned pro, you won't want to miss out on this exciting adventure!

The benefits of using AI for your business

Enhanced Efficiency through Automation. Imagine a world where you and your team can use your valuable brain power to create, strategize, and maximize relationships rather than complete repetitive tasks all day. That's what AI can do for your business. AI algorithms streamline job responsibilities such as data entry, customer service, and other tedious jobs. By automating routine and time-consuming tasks, AI can free up valuable time and resources, allowing you to focus on more important things that require more strategic thinking, planning, and creativity, such as developing new products or services and brand development. This automation means drastically minimizing wasting time on mundane, boring tasks. AI algorithms can automate those pesky tasks that take up too much of your and your employees' valuable time, allowing for more productivity.

Smarter Business Decisions. Gone are the days of guessing and hoping for the best. AI can handle vast amounts of information instantaneously, providing businesses with meaningful insights that can help make informed decisions, such as what time of year is optimal to release a new product or service or the best times to run sales or promotions. Whether you're trying to increase sales, streamline operations, or find new growth opportunities, AI has you covered. And that's not all. AI algorithms can also analyze historical data to make predictions about future trends and patterns, giving you the upper hand in the game of business. You'll always be one step ahead, knowing what the future holds.

With these valuable insights and predictions, AI can pinpoint the best solutions to help your business reach new heights. AI gives you an edge to better plan, strategize and implement business ideas. It can analyze customer feedback, such as reviews and social media posts, to understand customers' opinions and emotions. This function can help businesses make informed choices about product development, marketing, and customer support, improving customer satisfaction and brand reputation.

Improved Customer Service. We know that without customers, there is no business. Your customers deserve the best, and AI can help you provide just that. AI-supported chat agents, self-service portals, and virtual assistants can provide 24/7 support, improving the customer experience and increasing customer satisfaction.

Businesses use AI to enhance customer interactions and improve the customer experience. In addition, while AI handles your basic customer inquiries, your team can attend to more complex issues. AI can also analyze customer information to provide personalized product recommendations, making your customers feel heard and valued. This level of attentiveness can lead to increased customer loyalty and repeat business. AI is considered the superhero of client service for this reason, and it provides more efficient and personalized support that will take your customer experience to the next level. AI can also help improve customer interactions by providing quick and accurate answers to customer questions. Companies can use AI virtual agents to respond to more straightforward customer inquiries, making more time available for human representatives to handle more complex issues.

Predictive Analytics. Looking into the future has never been easier! AI algorithms can analyze historical data to make predictions about future trends and patterns, allowing businesses to be more proactive and avoid potential problems. AI predictive analytics uses vast amounts of information that provide valuable insights into consumer habits and market trends so that you can pinpoint where your industry is headed and what your next steps should be.

Cost Savings. Running a business can be expensive, but AI can help you save big bucks. Automating manual workflows and reducing the need to hire manual labor can result in significant cost savings. Lower labor costs mean less money allocated towards paying for salaries and benefits and more money added to your bottom line. Additionally, AI can audit and analyze data to identify areas where your company can reduce costs, such as energy usage or supply chain optimization. This benefit makes AI the perfect budgeting tool to ensure that you are maximizing your company's profits. AI can help you reduce waste and maximize efficiency. More innovative energy management with AI, for example, will optimize energy consumption, lower carbon footprint, and reduce waste. That means businesses cut costs and save money.

Innovation. AI enables businesses to develop new products and services and to create new business models that meet changing customer needs. With AI's unique capabilities, businesses are becoming more innovative. AI can help companies overcome obstacles and find new and creative solutions to their challenges. Ultimately, AI is a powerful tool for discovering new ways of doing business while continuing to evolve. It can help companies adapt more quickly to customers' needs, allowing them to stay on the cutting edge in today's fast-paced business world. This technology has made it possible for teams to work remotely and collaborate with partners and customers from all over the world, increasing their ability to develop new ideas and bring them to market more quickly. Overall, AI allows businesses to explore new ideas, experiment with new approaches, and find new and creative solutions to meet the changing needs of their customers.

Accuracy at its Best. We all make mistakes, but AI algorithms can detect patterns and identify anomalies with greater accuracy than humans, reducing the risk of errors and improving overall accuracy. With AI, you can rest assured that your business will always run like a well-oiled machine. AI algorithms can automate job duties, reducing the risk of human error, resulting in less time spent correcting employee errors and minimizing the financial and legal risks that some employees' mistakes can leave behind. This greater accuracy provides increased reliability and assurance that you can confidently operate your company without the fear of making costly mistakes.

Streamlined Processes. One of the most significant ways AI can help streamline business workflows is by automating tedious tasks. Companies can use AI to process invoices automatically, freeing up time for employees to prioritize more important tasks. Companies can also streamline HR workflows by automating resume sorting and identifying the most qualified candidates, saving HR employees time and ensuring that the best candidates are being considered.

Enhancing Marketing Efforts. AI can be used to improve marketing efforts, and companies can use it to analyze customer information to predict which products and services they might be interested in buying. More personalization of their marketing efforts results in increased conversions and less wasted time. AI takes the guesswork out of marketing by making more precisely targeted ads and helping create a marketing strategy most effective for the intended target market.

AI has proven to be the ultimate business sidekick, providing benefits such as increased efficiency and productivity, an enhanced customer experience, better predictive analytics, cost savings, and improved accuracy. Every business in any industry can use AI to scale faster. Why not yours?

Common AI applications and what they can do for your business

Let's dive into the world of AI and check out some of its innovative applications that can make a company's life easier and more efficient!

Customer Service Bots. These AI-driven robots are a company's secret weapon for providing quick and accurate customer support. They can handle a large volume of queries and complaints, simple requests like answering FAQs, and directing customers to resources 24/7 without taking a break! They can also be programmed to resolve the most common issues and even direct customers to the correct department to get further assistance. Customer support automation can be used in any size of business, allowing human agents to provide more personalized service when needed.

Sales and Marketing Analysis. AI can help companies analyze consumer behavior, such as buying patterns, to predict their purchasing interests. These insights can then be used to create personalized marketing campaigns and product recommendations, leading to increased sales and customer satisfaction. AI tools can analyze large quantities of data and provide insights into consumer trends, helping companies pinpoint issues with more accuracy. Have you ever abandoned your shopping cart at an online retailer and later received an email reminding you to complete your purchase? Or, maybe you've received a discount to incentivize you to come back and shop again after a long time has passed. This is AI at work! AI is an extra set of eyes to monitor customer activity and follows up so that human employees don't have to.

Supply Chain Management. AI can help make companies more efficient by optimizing the supply chain process by predicting demand, improving inventory management, and reducing waste.

It can also help with route optimization, reducing transportation costs, delivery times, and carbon footprint.

Fraud Detection. AI can provide an added layer of protection by helping companies identify and prevent fraudulent activities such as credit card fraud, insurance fraud, and cyber attacks. It can analyze vast amounts of data and detect patterns that indicate fraudulent activity, helping companies save time and money. AI algorithms can detect and respond to fraudulent activity without delays, protecting small businesses against financial losses and avoiding potential lawsuits due to data breaches. AI can enhance security by ensuring the safety of their information and systems.

HR Management. AI can help with recruitment by sorting through resumes and job applications, identifying the most qualified candidates, and scheduling interviews. It can also help with employee retention by analyzing employee satisfaction and engagement data and recommending improvements. AI algorithms can streamline the recruitment process, thus reducing the time and effort required to find and hire new employees.

Personalized Healthcare. AI can help healthcare companies provide customized care by analyzing patient information, including medical history and genetic information, to create individualized treatment plans. It can also help doctors and researchers by giving an up-to-the-minute analysis of patient data.

Inventory Management. AI algorithms can optimize inventory management. AI algorithms can predict when stock levels are low, allowing small businesses to order new products before they run out and eliminating the lag time between restocking. Out-of-stock products equal missed sales opportunities.

Financial Forecasting and Analysis. AI can help companies predict future financial performance and plan ahead for spending, investing, and growth. Here are some examples of what AI can do with financial forecasting and analysis:

- *Forecasting sales:* It can look at past sales data to predict how much a company will sell in the future, helping them plan for inventory and production.
- *Budgeting:* It can help companies predict their future expenses and income, so they can create a budget that makes sense for their business.
- *Investing:* It can analyze information to help companies decide where to invest their money, such as in new markets or product lines.
- *Managing risk:* It can help companies identify financial risks, such as market and industry changes, to avoid potential problems.
- *Streamlining accounting:* It can digitize routine responsibilities like bookkeeping, making it easier to keep track of financial data and create accurate forecasts.
- *Predictive maintenance*: It can predict when equipment and systems may fail, allowing small businesses to perform maintenance before issues arise. This foresight can reduce downtime, increase efficiency, and lower maintenance costs.
- *Market analysis and forecasting*: It can analyze vast amounts of information to identify market trends and forecast future demand. With this insight, companies can better strategize on product development, pricing, and marketing.

These are just some examples of the many ways companies can leverage AI to improve their operations and scale up; its uses are limitless! With AI, companies can save time and money, improve customer satisfaction, and stay ahead of the competition. As technology evolves and advances, the opportunities to leverage AI and technology to become more efficient and profitable will only continue to increase. So, if you're not already on the AI bandwagon, it's time to hop on and see what it can do for you!

| 3 |

AI Use Cases for Business Growth

> "The gradual platformization of AI is very interesting to me. The efforts by Google, Amazon, Salesforce — they're bringing AI down to a level of not needing to be an expert to use it. ... I think the day that any good software engineer can program AI will be the day it really proliferates." — **Kai-Fu Lee, Chairman and CEO of Sinovation Ventures**

In the rapidly evolving business landscape, making data-driven decisions is crucial for growth and success. Below are some illuminating case studies on notable brands that have effectively harnessed the power of data-driven decision-making to achieve significant advantages in the marketplace. These are just a few real-life examples of companies that have successfully used AI to scale and improve their operations:

Amazon has used AI to personalize the shopping experience for its customers, providing recommendations based on their purchase history and other information. Amazon's use of AI algorithms to analyze customer feedback has made it easier to drive more sales and get repeat business. Additionally, Amazon uses AI to optimize its supply chain and improve delivery times, reducing costs and improving efficiency, although not perfectly.

Netflix uses AI to personalize its recommendations for its users, suggesting streaming content based on their viewing history and other data. It also uses AI to analyze customer responses to decide which shows to produce and which to cancel.

Google uses AI to power its search engine and provide relevant results for users. It has also used AI to improve the efficiency of its data centers, reducing energy consumption and costs. AI is also responsible for the creation of the very popular Google Lens and Google Translate.

Walmart, one of the world's largest retailers, has been using AI algorithms to analyze customer information, optimize pricing strategies, and improve its supply chain and logistics operations. The retailer also uses AI to track customers' online search queries and sales data to better predicted consumer demand. The result is improved customer satisfaction, reduced costs, and increased profits.

Uber, a leading ride-sharing company, uses AI to make data-driven pricing, routing, and customer support decisions. By analyzing traffic information data with AI, the company has been able to estimate arrival times, adjust pricing dynamics on demand, optimize routing algorithms, reduce wait times, find the best match between drivers and riders, and improve safety.

These examples demonstrate how AI is transforming the way companies have better judgment by analyzing vast amounts of data, uncovering hidden patterns and trends, and reducing the risk of human error. Using this information to guide them forward, companies can achieve their goals, improve customer satisfaction, reduce costs, increase profits, and much more.

Small businesses have also leveraged AI technology in their services to their customers. AI is not limited to being used as a tool in your business - but also as a business! AI can be the actual product or service that you sell to customers to make their lives easier. Here are some examples.

H2O.ai is a small AI software company that provides businesses with machine learning and predictive analytics tools. Its AI software helps companies implement cyber-threat detection, predict cash demand, evaluate creditworthiness in underwriting, and a whole host of innovative solutions. In addition to being a user of its software, H2O.ai has helped small companies to analyze data and make better decisions, driving growth and improving operations.

Ascent is a small AI B2B company that helps organizations monitor changes in the financial industry's regulations so that companies can remain compliant and avoid legal trouble. As laws change, Ascent's AI software can quickly and easily detect the change and alert compliance managers, corporate attorneys, or other responsible parties.

Behavioral Signals is an AI-powered company that uses AI to analyze human behavior. Its advanced algorithms and machine learning techniques extract insights from various behavioral signals such as facial expressions, body language, and tone of voice. They use this information to understand emotions, motivations, and intentions. These insights can be used for various purposes, such as improving customer experiences, optimizing sales workflows, and reducing risk in financial transactions. In short, Behavioral Signals uses AI to gain a deeper understanding of human behavior and uses that information to benefit its clients.

Some notable women entrepreneurs have not only supercharged their brilliance by using AI to improve their operations and profitability but also created AI platforms to benefit other companies.

Some of these women have taken center stage in the AI tech sector and are pioneers in this space.

Penny Herscher is the CEO of *FirstRain,* an AI-enabled business intelligence platform. FirstRain uses AI to provide businesses with valuable insights and information. They use advanced algorithms and machine learning techniques to analyze an abundance of information from sources like company websites, news articles, social media, and more. The AI system then organizes this information and presents it in an easy-to-understand format, allowing businesses to stay on top of trends, understand their customers better, and uses this information to change how they do business.

Kriti Sharma is the Founder of *AI for Good,* an AI company that uses machine learning to solve social and environmental problems. She has used AI to improve the lives of people and communities worldwide and expand her impact. AI For applies artificial intelligence technologies to help address issues such as climate change, healthcare, education, and more. The company uses machine learning algorithms, extensive data analysis, and other AI techniques to create solutions that positively impact society and the planet. Their goal is to harness the power of AI to make the world a better place and to use technology for the greater good.

Dr. Ayesha Khanna is the CEO and co-founder of *ADDO AI,* a consultancy firm that offers AI solutions to businesses and governments. Its expertise in machine learning, data engineering, and data governance delivers its clients a strategic advantage by unleashing the power of data as a competitive edge. ADDO AI creates intelligent data platforms and automated processes that tackle the biggest challenges and arms its clients with a competitive edge in the global market by combing through various data sources and extracting undiscovered insights from them.

These examples show how women entrepreneurs are slowly beginning to utilize AI to drive growth, improve their businesses, and create businesses in the AI space. By leveraging the power of AI, women entrepreneurs can improve their operations, reach new customers, and achieve their goals, helping them to succeed in today's fast-paced business environment.

Nonprofits that leverage AI to scale

Nonprofit organizations are at the forefront of creating positive change in society. They use innovative and effective solutions to achieve their goals, and AI is one of the most powerful tools. Here are some specific examples of how nonprofits have used AI to achieve success and grow their organizations.

The *American Cancer Society* uses AI in a variety of ways to understand cancer better and improve patient outcomes. They employ AI algorithms to analyze vast amounts of medical data and identify patterns in patient information, which helps them develop new and more effective treatment plans. Additionally, the organization has implemented AI-enabled virtual assistants that provide patients with round-the-clock support and information about cancer treatments. These AI virtual assistants are trained on a vast database of information and can answer questions, provide advice, and connect patients with the right resources.

UN World Food Programme, one of the largest humanitarian organizations that tackles world hunger, leverages AI to distribute food more efficiently and reach people in need faster. They use AI algorithms to analyze satellite imagery, which helps them identify areas where food aid is most needed and to determine accessibility for delivery. Additionally, the organization uses AI-supported drones to deliver food supplies to remote areas where traditional delivery methods are not feasible. These drones are programmed with AI algorithms that help them navigate and deliver supplies safely and efficiently.

The *Wildlife Conservation Society* uses AI to track and monitor wildlife populations, which facilitates their conservation efforts. They use AI-powered cameras to collect images of wildlife, which are then analyzed by AI algorithms to determine the size and health of different species populations. The organization also uses AI to predict where wildlife will move next, which helps them protect habitats and prevent species from becoming endangered.

UNICEF uses AI to improve its work in several ways. The organization employs AI algorithms to analyze children's health and education information to allocate resources better. Additionally, UNICEF has implemented AI-driven chat agents as part of their *Safer Chatbots* project to provide information and support to needy children. These AI virtual assistants are designed to understand children's questions and provide age-appropriate responses to them when they are in distress, need to reach a trained professional for help, and seek discreet assistance to avoid danger.

These examples show how companies of all sizes, industries, and sectors can use AI to improve operations, reduce costs, solve social problems, and succeed in today's fast-paced and competitive business environment. By leveraging the power of AI, organizations can stay ahead of the curve, avoid becoming obsolete, and compete with larger companies, all while making a more significant impact on society and helping more people in need. AI is a powerful tool that has the potential to revolutionize the nonprofit sector as this technology is more widely adopted, and these success stories demonstrate its potential to create a positive impact in the world.

AI tools designed specifically for nonprofits

As shown in the previous examples, the use of AI technology is not exclusive to for-profit companies. In fact, more and more AI tools are being developed to help nonprofit organizations scale their organizations, become more efficient with their resources, stay compliant with proper governance, and reach their donor goals. There are a number of AI tools specifically designed for nonprofits, and here are just a few popular ones.

1. *Twilio.org* is an AI-enabled nonprofit that helps other non-profits increase their efficiency and impact. It maximizes fundraising efforts with personalized campaigns that target each unique supporter's interests and preferences. The organization's AI systems generate attention and keep donors engaged with relevant and timely communications across channels like text messaging, phone calls, virtual chat agents, and email. At its core Twilio helps nonprofits personalize every interaction in a digital-first world using real-time data and design visually appealing emails with intuitive segmentation and track performance with built-in analytics.

2. *Keela* is a fundraising CRM designed for nonprofits. It offers powerful tools to help organizations reach new audiences and build a community of supporters. One of its features, Donor Readiness, is an AI-based tool that calculates a donor's likelihood of donating within the next two weeks by analyzing various factors such as giving history, contact interactions, and even variables like contact location and weather. This tool helps organizations maximize their fundraising potential by determining the best time to reach out to donors, improving their chances of giving. With Keela's Donor Readiness,

nonprofits can boost their fundraising results and build life-long relationships with their supporters.

3. *Dataro.io* uses AI to help nonprofits raise more funds and reach more donors. Its AI technology helps organizations filter donors based on propensity scores, allowing them to create personalized messages for each unique segment. By targeting donors with a higher likelihood of giving, organizations can significantly increase conversion rates and raise more funds. Dataro also helps organizations identify donors with the highest propensity to set up recurring donations or reactivate lapsed ones. Additionally, its AI software predicts which donors are at risk of churning and gives organizations insights to strengthen their stewardship strategies, as well as help with major gift fundraising strategies by identifying supporters most likely to respond favorably to outreach.

Case studies: a tale of two nonprofits

Parkinson's UK

Parkinson's UK utilized Dataro's AI-powered supporter predictions to refine their appeal with pinpoint precision. During their seasonal direct mail campaign, Dataro's machine learning algorithm analyzed their historical fundraising data to create a focused list of donors who were most likely to contribute to the campaign. The results of this AI-driven approach were compared to those of the organization's own donor selection strategy. Dataro's predictions achieved a response rate of over 14%, while the traditional segmentation approach yielded just under 9%. Focusing on donors identified through machine learning led to a potential revenue increase of 23% and resulted in over 2,800 individual gifts, despite working with a smaller list of potential donors. By leveraging machine learning, Parkinson's UK conducted a more effective campaign that raised more funds for its mission while reducing printing and mailing expenses.

Greenpeace Australia

Greenpeace Australia partnered with Dataro to enhance their fundraising efforts by using AI in two ways: direct mail donation appeals and reducing churn in their regular giving program. Dataro analyzed Greenpeace's entire fundraising, engagement, and communication data to assign a propensity score to each donor, indicating their probability of giving through a direct mail appeal. The AI-generated list of donors who were most likely to give after receiving an appeal was compared with Greenpeace's traditional RFM segmentation list. The campaign was executed using both lists, and the AI-generated list outperformed the traditional list. Additionally, Dataro's machine learning tools were used to identify regular givers who were at risk of churning, and these donors were contacted with personal "thank you" calls, resulting in retaining 64 donors who would have otherwise lapsed and saving an estimated $23,040 for the organization. By leveraging AI-driven fundraising tools, Greenpeace was able to increase donations and reduce costs.

By leveraging AI, nonprofits can become more effective at making their impact in society and focus more on helping more people in need rather than on tedious tasks that can be automated. AI technology offers significant advantages for nonprofits, including improved fundraising tactics and operational efficiencies. With the support of AI tools, organizations can raise more funds, target donors more personally, reduce their cost of fundraising, and save time on data management and campaign planning. Implementing AI in fundraising operations can result in increased funds available for mission-driven activities rather than being spent on overhead expenses or the fundraising process itself. Although these amazing AI tools were made with nonprofit organizations in mind, many of the tools listed in *Chapter 8: 100 AI Business Tools to Scale Faster and Smarter* can also be used to improve and grow nonprofit organizations.

| 4 |

Limitations of AI

> "I think what makes AI different from other technologies is that it's going to bring humans and machines closer together. AI is sometimes incorrectly framed as machines replacing humans. It's not about machines replacing humans, but machines augmenting humans. Humans and machines have different relative strengths and weaknesses, and it's about the combination of these two that will allow human intents and business process to scale 10x, 100x, and beyond that in the coming years." – **Robin Bordoli, former CEO of Figure Eight**

AI is becoming more advanced and rapidly changing many aspects of conducting business. But despite its remarkable abilities, there are some things that AI can't do. In this chapter, we'll dive into the specifics of what AI can't do and why it's important to keep our expectations in check.

AI Can Replicate Aspects of Human Intellect, but Not Replace It

One of the limitations of AI is that it can only mimic human intelligence. While AI can be programmed to do certain things, it doesn't have emotions, creativity, or the ability to think critically as we do. AI can analyze lots of information and give us recommendations, but it can't understand what it's like to be a person.

AI Can't Make Decisions on Its Own

Another thing to remember is that AI can't make decisions without us. All AI systems are created and programmed by humans, and their decision-making skills are limited by the data they've been given and the algorithms they use. This means that AI is only as good as the information it gets and can only provide answers based on that. For instance, AI in healthcare needs to be fed accurate and comprehensive information in order to give accurate diagnoses. If the data is missing or incorrect, diagnoses will also be wrong.

AI Can't Replace Human Judgment

It's important to know that AI can only give us information and recommendations but cannot replace our human judgment. AI doesn't have the wherewithal to weigh the ethical, social, or moral impact of its decisions, and it can't come up with answers or solutions based on intuition or common sense. So, even with AI involved, humans still need to be a part of the process of reaching the desired outcome. For instance, AI in criminal justice must be guided by human judgment to ensure fairness and impartiality. AI can't make a choice based on the complexity of human behavior.

AI Can't Be Trusted Without Us Monitoring It

To make sure that AI is trustworthy, we need to keep an eye on it. AI systems can be distorted due to the programming inputs of biased data, make mistakes, or have unintended consequences, just like any other technology. So, it's crucial to have proper oversight and regulations to ensure AI is being used responsibly and ethically. One way to explain this is that AI in financial markets needs to be regulated to prevent unethical practices like market manipulation.

Bias and Discrimination

This is a sensitive topic that is discussed a lot among experts in the AI space. One of the limitations of AI is its tendency to perpetuate partiality and discrimination. This limitation is because AI is only as good as the data it's trained on, and if it contains biases, then the AI outputs will reflect those biases. To illustrate, if an AI system is trained on a dataset of job applicants that has predominately men, it may not be able to recognize the skills and qualifications of female applicants. Similarly, suppose an AI system is trained on a dataset of criminal offenders that contains mostly people of color. In that case, it may be more likely to label people of color as potential criminals, even if they haven't committed any crimes. Therefore, AI used in law enforcement must be monitored to make sure it's not being used for racial profiling or other forms of discrimination. To combat unfairness and discrimination in AI, it's imperative to see that the information used to train AI systems is diverse and representative of the population it's meant to serve.

Limited Understanding of the Real World

Another limitation of AI is its inability to understand the real world. AI systems are designed to make predictions and conclusions based on patterns and trends in data, but they can only do so based on the information they have. For instance, if an AI system is trained on information about traffic patterns, it won't be able to accurately predict traffic conditions on a rainy day if it has been programmed with limited information about how rain affects traffic. AI systems can only process information based on the exactness of the data and instructions it receives.

Limited Creativity and Problem-solving Ability

AI systems are great at recognizing patterns and making predictions, but these systems are not very good at creative problem-solving. Why? Because AI systems are designed to follow rules and algorithms, and they do not have the ability to think outside the box or come up with new and innovative solutions. To illustrate, an AI system designed to play a game like chess may be very good at following the rules and making strategic moves. Still, it won't be able to conceptualize new and creative strategies that humans have never seen before. To overcome this limitation, it's critical for AI systems to become more flexible and adaptive and that they are able to learn and evolve over time.

Lack of Common Sense

The ability to understand things without explanation is sometimes referred to as common sense. Another limitation of AI is its lack of common sense. AI systems cannot understand the world and how things work unless it is explained to them by humans in the form of algorithms. This lack of common sense makes it challenging for AI to provide answers in situations that are both unpredictable and complex. To further explain, an AI system designed to play a video game may be able to recognize patterns and produce outcomes based on those patterns, but it may not understand the game's basic rules or how the game world works. AI technology may never evolve enough to possess the depth of understanding required for human common sense and judgment, which is why its intelligence is artificial.

Limitations in Natural Language Processing

Natural language processing (NLP) is one of the most challenging areas of AI, and it still has many limitations. NLP systems are designed to understand and generate human language. However, it still has a long way to go before it can truly understand the complexity and nuances of human language. For instance, NLP systems may have trouble understanding sarcasm or humor and have difficulty recognizing and processing different dialects and accents.

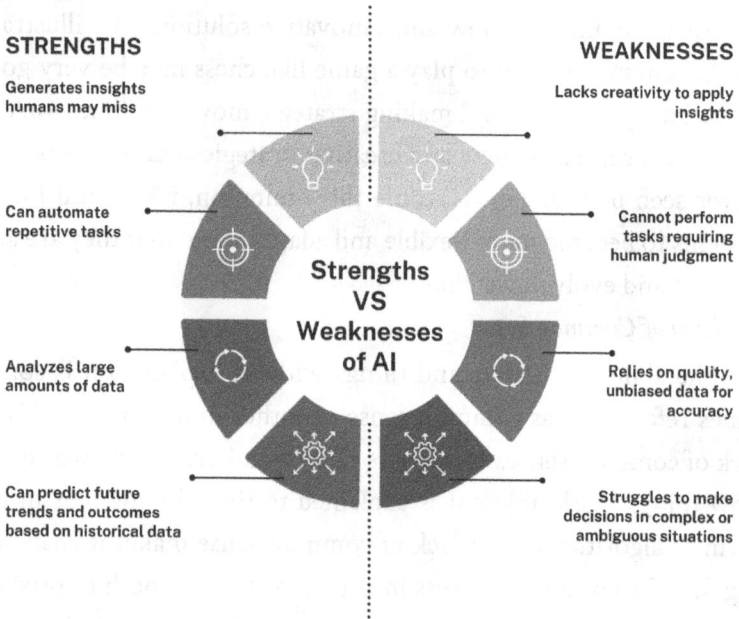

STRENGTHS

Generates insights humans may miss

Can automate repetitive tasks

Analyzes large amounts of data

Can predict future trends and outcomes based on historical data

WEAKNESSES

Lacks creativity to apply insights

Cannot perform tasks requiring human judgment

Relies on quality, unbiased data for accuracy

Struggles to make decisions in complex or ambiguous situations

Strengths VS Weaknesses of AI

Figure 4.1 Strengths VS weaknesses of AI

Overall, the correlations between its strengths and weaknesses highlight the need for businesses to carefully consider the potential benefits and drawbacks of AI before implementing it. They must determine if they have the necessary data, infrastructure, and ethical frameworks in place to support the use of AI and maximize its potential benefits while minimizing its limitations. If the benefits outweigh the drawbacks, then it is a sensible decision to implement the AI solution.

AI has the potential to become even more impressive, but there are some things it just can't do. It can only mimic but it cannot replicate human intelligence, make decisions independently, replace human judgment, or be trusted without humans monitoring it or telling it what to do. You still have to think for AI in terms of giving it the correct input so that you can get the correct output.

By understanding the limitations of AI, you can use it with the right expectations and awareness of how it can supercharge your brilliance but not replace it!

| 5 |

"AI, You're Hired!" Employing the Right AI Tools for Your Business

> *"Artificial intelligence is one of the most profound things we're working on as humanity. It is more profound than fire or electricity."* **– Sundar Pichai, CEO of Alphabet (Google)**

Leaders of organizations are constantly seeking new ways to leverage technology to work smarter as the world becomes more digital than before. However, the process of implementing AI can be a daunting one for some. From choosing the right tools to understanding how to integrate them into existing workflows, there are many factors to consider. That's where this guide comes in. We'll explore some key considerations for business leaders who are looking to leverage the right AI tools in their operations.

How to pinpoint areas where AI can make a difference

As extraordinary as AI tools are, many companies may need help knowing where to start or how to best utilize AI in their business operations. There are steps that companies can take to identify precisely what areas where AI is needed and how they can make a substantial difference and maximize its benefits.

1. *Analyze Your Data*: The first step to utilizing AI in your company is assessing the information you currently collect and process. Ask yourself the following questions:

   - Is a large amount of data being handled manually?
   - Is the data messy and difficult to make sense of?
   - Do you want to extract useful insights from your data but don't know how?

If the answer to any of these questions is yes, then you can use AI to clean, organize, and make sense of your company's data, making it easier to use. AI can also identify patterns and trends within it that humans may not notice quickly. This capability can help companies gain new insights that improve aspects of business going forward.

2. *Identify Repetitive Tasks*: AI can automate recurring job tasks that are time-consuming and prone to human error. Examples of these tasks may include manual resume reviews, data entry, or even customer inquiries. By automating these tasks, companies can free up valuable time and resources, allowing employees to direct attention to more strategic and creative tasks.

3. *Analyze Consumer Behavior*: AI can analyze customer information and predict their future needs and preferences, helping companies personalize their products and services to meet customer needs better and with less human intervention. Precisely anticipating

what customers want reduces time wasted and money on trial and error.

4. Improve Operations: AI can instantaneously analyze large amounts of data, quickly uncover problem areas, and know what course of action to take to improve best practices in specific areas such as employee hiring decisions, budgeting, inventory management, and regulatory compliance.

Companies can improve their operations, reduce costs, and increase profits by utilizing AI to automate routine job responsibilities and make data-driven decisions. It's important to reiterate that while AI can bring many benefits, it should not be seen as a replacement for human decision-making but rather as a tool to support and enhance it.

Implementing AI can bring numerous benefits, but it's necessary to understand how best to utilize it. Following the steps outlined in this guide can help you identify areas where AI can make a difference and lead to more efficient operations and increased profits.

The AI toolkit: Choosing the right AI tools and solutions for your business

Finding the right tools and resources for AI is critical to solving the right problems and achieving the goals you've set to scale your company. There are many AI-powered analytics platforms, so finding one that fits your needs and budget is essential. However, before choosing the right AI tools, it's important to be sure that you have accurate and relevant information to feed into the algorithms. As the saying goes, "garbage in, garbage out!" In this context, the correct data integration with AI tools plays a vital role in scaling up your business.

Consider factors such as the size of your business, the level of complexity you're comfortable with, and the amount of information you need to analyze. Once you have the right tools and resources, you can then select the proper AI-driven tools that will help you achieve the goals that you have set for your organization. Here are some recommended action items.

Determine Goals. Before companies can choose the right AI tools and solutions, they must first determine their AI needs, which means looking at their current workflows and processes and identifying areas where AI can help. What do you want to achieve with AI? Do you want to improve efficiency, reduce costs, or automate tasks? Identifying your business goals will help you to choose the right AI tools and solutions for your needs.

Determine your Budget. Some AI tools can be expensive, so consider your budget when choosing an AI platform or solution. You may need to invest in hardware, software, and personnel to implement AI.

Research AI Solutions. Once a company knows its AI needs and budget, it's time to do some research. There are many AI solutions available, from open-source software to commercial products. Companies can start by checking out AI industry blogs, watching YouTube videos, attending AI conferences, and researching other companies that have implemented AI solutions.

Check your Data: The success of AI depends on the quality of the data it uses. Before choosing an AI tool or solution, it's critical to have high-quality information to feed the algorithms that are relevant to the results you desire to achieve. If you don't have high-quality data, you may need to invest in data collection and cleaning before implementing AI.

Evaluate AI Solutions. Next, companies should evaluate the different AI solutions to determine which fits best. This means looking at factors like pricing, ease of use, integration with existing systems, and required maintenance. Companies should also consider the level of technical support available and the quality of the AI solution's algorithms. Another consideration is scalability. As your organization grows, your AI needs may change. Before choosing an AI tool or solution, it's crucial to make sure it's scalable and can grow with your business.

Build an AI Team. Building an AI toolkit might require a multi-person team, depending on your organization's size. Companies should build a team of experts to help choose and implement the right AI solutions. This team should include people with various skill sets, from data scientists to developers to business analysts. If you don't have the right skills and expertise in-house, you may need to invest in training current staff, hiring new staff, or outsourcing talent.

Test and Refine. Finally, companies should test their AI solutions to see if they work as expected. If there are any issues, they should be refined and improved. Companies should also regularly review their AI solutions to make sure they're still meeting their AI needs.

Using AI-powered analytics to gain a competitive advantage

AI-powered analytics is like having a superpower for your business. It allows you to analyze vast amounts of information in seconds and reveal insights not easily detected by the human eye. It will become clear precisely what areas of your operations you need to optimize, such as which products or services require more marketing, which workflows are inefficient, where you're losing money, and where you can save money.

These analytics can track the performance of your products and identify which ones are selling well and which need improvement. It can also help determine which promotions are most effective. With this information, you can improve your purchasing and inventory needs to precisely know which products to stock, when to order new merchandise, and which promotions to run.

With analytics optimizing your operations, you can improve with real-time insights to track your expenses and identify areas where you can cut costs. It can monitor your supply chain by better tracking and managing inventory levels, delivery times, and shipping costs. This information makes you aware of which suppliers and shipping companies are most suitable to use.

Using the analytics gathered by AI makes your company faster, more efficient, and more effective than your competitors. With the time and money saved by automating processes with AI, more time and money can be allocated towards expanding into new markets and innovating by developing new products and services that launch ahead of competitors. The advantage AI gives organizations can help to gain market share and become the preferred brand by more customers.

A special mention about ChatGPT

"Artificial intelligence would be the ultimate version of Google. The ultimate search engine that would understand everything on the web. It would understand exactly what you wanted, and it would give you the right thing. – Larry Page

If you're anything like me, you became even more intrigued about artificial intelligence when ChatGPT debuted in November 2022. It has had much of the world in a frenzy upon its introduction!

This AI tool deserves its own section because its uses to improve your business are nearly endless.

ChatGPT is like the Google search engine on steroids. It is an AI language model developed by *OpenAI and* a highly sophisticated technology that is changing the game for companies, consultants, content creators, gig workers, professionals, and alike in every industry. It is a cutting-edge AI tool that can perform a wide range of language-based tasks, making it a versatile solution for various industries. Here are the 20 popular uses of ChatGPT for businesses and professionals.

1. Content creation - Use ChatGPT to generate blog posts, social media posts, articles, product descriptions, coaching packages, website content, professional biographies, ad/sales copy, catchy taglines and slogans, emails, summarize lengthy content, speeches, sales scripts, song lyrics, screenplays, and more.

2. Agreements and contracts - It is always advisable to have a licensed attorney review contracts created by ChatGPT for accuracy. However, paying an attorney to review an existing contract rather than creating an entirely new one might save you lots of time and money!

3. Excel formulas - To illustrate, ask it to create lookup and reference formulas for Excel to search for and retrieve specific data from a range of cells or a table. Or, ask it to create statistical formulas to calculate various statistical values, such as averages, standard deviations, and correlations.

4. Product launch schedule - Input the launch date and key milestones and assign task completion dates, and ChatGPT can instantly design or adjust a product launch schedule.

5. Project timeline - Input the project's scope, list the tasks, estimate the duration, and assign resources to each task, such as personnel or materials, to create a project timeline.

6. Write code - ChatGPT can write Python, JavaScript, HTML, SQL code, and more.

7. Develop custom AI virtual assistants - ChatGPT can provide customer service, automation of frequently asked questions, lead generation, and more.

8. Explanations for complex subjects – When entering prompts, ask it anything from Astrophysics to Pythagoras' Theorem to Zaitsev's Rule.

9. Translation and language processing - Use ChatGPT to translate text quickly, summarize long documents, and perform other language-related tasks.

10. Academic/scholarly research – ChatGPT can conduct the research from existing published works ranging from articles to full papers. Be sure to credit the original author's work if known!

11. Virtual writing and editing assistance - Use ChatGPT to assist with writing and editing, including proofreading and grammar checking.

12. Event planning of virtual and live events – Enter as many parameters for planning your event as you know, such as the type of event, dates, time frames, format, number of participants, etc., and ChatGPT can create an entire event plan.

13. Personal shopping and styling ideas - This includes recommendations of clothing and accessories based on individual style and body type

14. Grant writing and proposals – Enter the known information from the RFP, and ChatGPT can create a grant proposal for you. The more specific the inputs, the more specific the finished product will be.

15. All types of reports, including financial reports – Monthly financial reports, fiscal year reporting, and more can be created with the help of ChatGPT.
16. Legal advice - Ask it questions about the law. (This is not a replacement for an attorney but serves as a great starting point!)
17. HR support to create company policy language - This includes full company policy manuals, employee training programs, code of conduct policies, and more.
18. Virtual marketing assistance - This can entail market research, generating content, adding SEO to content, custom marketing plans, and market research to gather, analyze and interpret lots of information to gain insights into consumer behavior and market trends.
19. Virtual data analysis for analyzing and interpreting an abundance of information and generating reports.
20. A search engine x 10,000!

This list is not even close to exhausting what ChatGPT can do to improve and streamline your operations and scale your business fast. With its advanced language processing capabilities, ChatGPT is a valuable tool for any professional looking to stay ahead of the curve and save lots of time and money while doing it.

| 6 |

Customer Engagement with a Twist of AI

> "*Humans need and want more time to interact with each other. I think AI coming about and replacing routine jobs is pushing us to do what we should be doing anyway: the creation of more humanistic service jobs.*" – **Kai-Fu Lee, Chairman and CEO of Sinovation Ventures**

Picture this: You're running a company and want to give your customers an experience they'll never forget. You want them to feel appreciated, valued, and heard. But, with so many customers to keep track of, it can be tough to ensure everyone gets the attention they deserve. That's where AI comes in! This tech-savvy sidekick can help you keep up with all your customers and provide them with an experience they won't forget.

Examples of AI solutions and strategies to keep your customers coming back

1. *Chatbots that provide 24/7 support.* Companies can use AI-enabled chat agents to provide customers with quick and accurate answers to their questions, even outside regular operational hours.

2. *Personalized product recommendations based on purchase history.* AI can suggest similar products they might be interested in and analyze a customer's purchase history to better understand their preferences.

3. *Instant sentiment analysis of customer feedback.* Companies can use AI to monitor customer feedback on social media and other platforms and respond to negative comments instantly, improving customer satisfaction and loyalty.

4. *Predictive maintenance for customer support.* AI can predict when a customer is likely to have a problem with a product, allowing customer support teams to proactively reach out and resolve the issue before it becomes a bigger problem.

5. *Virtual styling and try-on services for fashion and beauty brands.* Companies in the fashion and beauty industries can use AI to provide customers with virtual styling and try-on services, allowing them to preview products before making a purchase and creating a more engaging and personalized shopping experience.

6. *Voice-activated virtual assistants for customer support.* AI can allow for voice-activated virtual assistants to provide customers with hands-free support, making it easier for them to get answers to their questions.

7. *AI-powered product search and navigation.* Leveraging AI to provide customers with an intelligent product search experience

helps them find what they're looking for faster and more accurately.

8. *AI-enabled personal shopping assistants.* Companies can use AI to provide customers with personalized virtual shopping assistants, helping them find the products they're looking for and make the right purchase that suits them best.

9. *Predictive marketing based on consumer behavior.* AI can predict customer behavior and preferences, allowing companies to create targeted marketing campaigns that are more likely to engage customers.

10. *AI-supported virtual events and experiences.* Companies can use AI to create virtual events and experiences precisely tailored to customer preferences, providing customers with a more engaging and personalized experience.

These are just a few examples of how companies can use AI to enhance customer engagement. By using AI to provide customers with a more personalized, efficient, and engaging experience, companies can improve customer satisfaction and loyalty and increase sales. And while AI is handling routine jobs, human customer service representatives have extra time to attend to more complex issues that require personal interaction with customers, such as handling complaints or offering personalized assistance. In these instances, empathy, patience, and compassion, which are uniquely human traits, are needed and cannot be substituted with AI.

AI-powered personalization and its impact on customer experience

AI-powered personalization refers to using AI algorithms and machine learning to understand and predict habits, preferences, and needs. This information can then be used to create a more customized and engaging customer experience.

Think about this: Have you ever been shopping online and noticed that the website shows you recommendations based on your previous purchases? Or have you received a personalized email from a company promoting products that might interest you? That's AI-driven personalization in action, and it's almost like giving customers their own personal shopper with curated product ideas that they will likely be interested in buying.

By using AI to personalize the customer experience, companies can improve customer engagement, satisfaction, and loyalty. AI can recommend products or services that customers are more likely to be interested in buying or offer personalized promotions and discounts. Targeted marketing not only makes customers feel understood and valued, but it increases their likelihood of purchasing.

AI-empowered personalization is having a massive impact on customer experience. By using AI in this way, companies can create a more engaging and satisfying experience for its customers, making them happier and more loyal. So, take your customer experience to the next level with AI-supported personalization!

Finding the right balance between AI and human touch

Finding the perfect mix between AI and the human touch in customer service is vital to giving your customers a top-notch experience. Here's how to do it.

Know your customers. It's essential to get to know your customers to understand the level of service they expect. Some may want a personal touch, while others are cool with the convenience of an AI chatbot handling their needs. By knowing their preferences, you'll know when to bring in the bots and when to call in the human reinforcements.

Augment, but don't replace. AI can be a super helpful tool for handling routine tasks and simple inquiries, freeing human service representatives to tackle the tough stuff. But remember, AI is not the end-all, be-all solution. So before you think about firing your team, remember that you still need people. Use AI to improve the customer experience, but not to do away with the human touch.

Have humans on standby. No matter how smart AI gets, there will always be times when a human touch is necessary. That's why it's highly recommended to have human agents ready to jump in when the chatbot is in over its head or if the technology is temporarily out of service. Then, your customers will always be served regardless of technology mishaps.

Cross-train on AI. For the best customer experience, AI and human representatives must work together like a well-oiled machine. Train them to understand each other's strengths and limitations, and watch as they provide a seamless client service experience. It is vital to be able to switch between humans and AI to ensure a smooth process for the customer.

Keep Improving. It's crucial to continually evaluate and improve the balance between AI and the human touch in customer service. Therefore, it is ideal to keep track of customer feedback, use it to make changes, and keep striving for the perfect balance. That way, you can give your customers the best experience possible, regardless of their service preferences. By finding the sweet spot between AI and the human touch, companies can deliver fast, efficient, and personal service that meets their customers' needs.

| 7 |

Selling with AI Style

> "A lot was happening in A.I. But I also realized it wasn't clear what Salesforce's role in A.I. was. That's when we started acquiring quite a few artificial intelligence companies, maybe a dozen."— **Marc Benioff, CEO of Salesforce**

Getting prospective customers to buy from you is a science. Thankfully, AI can gather much of the information you need to sell with confidence. It's time to take your game to the next level with the help of AI. Imagine having a secret weapon that makes closing deals a breeze. No more tedious research or guesswork because AI has got you covered. So, when you are ready to wow your customers and close deals like a boss, the power of AI is at your service.

Using AI to Understand your target audience and their needs

Every company committed to having happy customers must be deliberate in finding out what their customers want and need. By doing so, companies can retain customers and keep them coming back for more. Now, how does AI come into play? Well, imagine having a group of super-intelligent robots at your disposal that can analyze massive amounts of data on your target market in a matter of seconds. That's what AI can do. With AI, companies can gather and analyze tons of information about them, including their likes, dislikes, and habits. This information is then used to create products and services that meet their needs.

The process of selling with AI involves using artificial intelligence technologies to analyze data and generate insights that can help sales teams identify potential customers, personalize communication, and close more deals. Here are the general steps involved in the process:

AI Sales Process

Data Collection	Data Analysis	Personalize Communication	Lead Scoring	Automation
1	2	3	4	5
PREPARE	IDENTIFY	CUSTOMIZE	RANK	IMPLEMENT
• Gather relevant data about customers and prospects. • Data can include demographics, preferences, behavior, and purchase history.	• AI algorithms analyze and identify patterns and trends in data. • Extract insights on customers' wants and needs and assess how they can be addressed.	• Use insights to create personalized communication with AI-powered tools. • Communications can include emails with targeted offers.	• AI algorithms score leads based on likelihood to buy rankings. • Use rankings to focus efforts on the most promising prospects.	• Automate routine tasks, such as sending emails, lead follow-up, and appointment scheduling. • Redirect sales efforts to close deals.

Figure 7.1 AI sales process

Selling with AI involves using data and insights generated by algorithms to create more personalized and effective sales strategies, resulting in increased sales and revenue. After implementing AI-powered sales strategies, it's important to continually monitor and analyze the results to improve performance. Sales teams can use AI-generated insights to optimize their strategies, refine their targeting, and improve their communication with customers. By constantly learning from the data, sales teams can stay ahead of the competition and continue to increase sales and revenue over time. The centerpiece to selling successfully with AI is understanding your target market.

Here are some ways companies use AI to understand their target audience

Social Media Monitoring. Companies use AI to monitor their target audience's activity on social media platforms. This automated monitoring includes tracking their likes, dislikes, comments, shares, and more. With this information, companies can better understand what interests their target audience the most, such as sales and specific products. Conversely, automated monitoring tracks complaints and negative comments, which are equally as helpful.

Customer Surveys. Businesses leverage AI to analyze customer surveys without delay. This way, they can quickly identify trends and areas that need improvement. And with the help of AI, they can also personalize their surveys, making them more relevant to each individual customer.

Sentiment Analysis. Companies rely on AI to analyze customer reviews and feedback to determine their sentiment. The feedback can help companies identify areas that need improvement, as well as areas where they're excelling. With this information, companies can make changes to the sales process to better meet their customers' needs.

AI is changing how companies understand their target audience, and it's all for the better. With AI, companies can gather and analyze more information faster and use it to create products and services that more precisely meet their customers' needs. And who doesn't love a product or service that was made just for them? It's like anticipating what customers want without having to ask. Sales and marketing teams can leverage AI's data to convert new customers and keep existing ones returning. Since AI makes it easier for companies to understand the needs of their target audience better, creating products and services that their customers will love. This is achieved by removing much of the trial and error and guesswork from trying to figure out what their customers want.

Utilizing AI to enhance marketing and sales

Targeting your audience with precision, examining customer purchasing history and preferences, and optimizing your marketing campaigns for maximum impact are all processes that would customarily require a significant amount of time. Now sophisticated AI tools can do all those tasks and more with record speed. This technology is here to improve how you approach marketing and sales, giving you an edge over the competition and helping you achieve your goals faster and more efficiently.

Selling can be challenging, especially in B2B, where the sales cycle can be longer and more complex. Customers don't always decide to buy right away and sometimes need a lot of guidance and assurance. So, sales representatives must build a trusted relationship with prospective buyers by speaking with them over the phone, meeting with them in person when necessary, and answering their questions during and after the sale. Even in the most complex sales process, AI is here to help. According to a study by Harvard Business Review, companies using AI in sales have seen an increase in leads by over 50%, a reduction in call time by 60-70%, and cost savings of 40-60% (Goergen, 2022).

So, how can AI enhance the sales cycle? Although it will not replace sales reps just yet, it can act as a helpful assistant to streamline some job duties, organize the sales workflow better, and provide great analytics to improve conversion rates. Automating tedious tasks can be done with AI, like data entry and meeting scheduling, or complicated jobs that don't require personal relationships, like sales forecasting. It also helps sales reps prioritize better by highlighting patterns in customer responses. The detailed analytics that AI provides on all communication between sales reps and potential clients, including emails, phone calls, and chats, can make selling less complicated and an overall better experience for the customer.

Marketing attracts buyers, and buyers result in sales. Although this seems like a relatively straightforward process, it can be nothing straightforward about it. However, with AI, companies can optimize their sales process. The detailed analytics that it can provide to sales reps, can improve their sales cycle, close deals more efficiently, and make customers happy. AI is becoming an indispensable part of selling for organizations, and they are using it for many of their sales activities.

Some of the primary sales activities and AI use cases are:

- More accurate sales forecasting with AI-powered demand forecasting
- Lead generation and lead scoring to help sales reps prioritize which leads have the best chance of closing
- Personalized ad copy and analytics to improve engagement rates
- AI to suggest the next best action for sales reps
- Automating mundane sales activities like data input and meeting setup

- AI-driven chat and email bots to help with outreach and make sales easier

AI is here to make sales more manageable and effective, so sales reps can attend to what they do best, building customer relationships.

Utilizing AI-powered marketing and sales tools to reach and connect with your target audience

AI can assist with selling your products on various digital channels, allowing you to reach customers in new and innovative ways. More reach can lead to more sales, and more sales can lead to more profits! With its cutting-edge technology, many AI tools can help you sell smarter, faster, and more successfully.

E-commerce platforms. AI can be integrated with popular e-commerce platforms like Amazon, Shopify, and Magento to enhance the customer experience. With AI, you can personalize product recommendations for each customer based on their purchase history and browsing behavior. Additionally, AI-enabled virtual assistants can handle customer inquiries and sales, providing customers with a quick and convenient way to get the information they need.

Social Media. Social media platforms like Facebook and Instagram have millions of users, and AI can help you tap into those audiences. AI algorithms can analyze follower behavior on these platforms and target them with personalized advertisements that are more likely to result in a sale.

Your Website. Your website is a critical component of your online presence, and AI can be used to optimize it for sales. AI algorithms can be integrated into your website to provide customers with personalized product recommendations, and AI virtual agents can handle customer inquiries and sales. Predictive analytics can also suggest additional products or services that offer more value and more easily upsell customers to increase revenue.

Mobile Apps. Many customers prefer to shop on the go, and AI can be integrated into mobile apps to reach these customers. Mobile apps take the shopping experience anywhere there's an internet connection. Since the vast majority of people take their smartphones with them everywhere they go, there's no need to miss out on any buying opportunities from customers who value this convenience.

Voice-enabled Devices. With the increasing popularity of voice assistants, AI can be used to sell products through voice-enabled devices like Amazon Alexa and Google Home. Customers can use their voices to search for products, compare prices, and make purchases, making shopping more accessible and more convenient.

AI can help boost sales and improve the customer experience when combined with human expertise and judgment. As always, AI works best when used in conjunction with human input, and effective AI-driven sales require access to high-quality information, a well-designed system, and ongoing optimization and improvement.

| 8 |

100 AI Business Tools to Scale Faster and Smarter

> *"I think about AI as a very powerful tool. What I'm most excited about is applying those tools to science and accelerating breakthroughs."* — **Demis Hassabis, co-founder and CEO of DeepMind**

Whether it's finance, marketing, sales, HR, or customer service, AI has got you covered! So many amazing tools can supercharge your brilliance and help you enhance almost everything your business does with lightning speed, less effort, and lower costs. Whether you are a freelance gig worker or a large multinational company, there is an AI tool that will work for you. It's also worth noting that AI is being used in medicine, real estate, entertainment, construction, and other industries that might not seem to have likely applications.

In addition to the tools already mentioned thus far, here are 100 more for you to leverage to scale faster. Most of these platforms can be used by any sized organization, even small ones that need free or budget-friendly AI platforms to scale. Are you ready to accelerate your business growth to the next level with the help of some game-changing AI software tools? Then buckle up, grab your laptop or tablet, and get ready to journey into the world of AI tools!

100 AI tools that grow your business fast!

1. *Acrolinx* is an all-in-one platform that helps businesses set and customize content goals, generates analytics for agile content workflows, and promotes effortless alignment across teams. It provides all the necessary features to optimize content from a single platform.

2. *Act-On* offers a marketing automation platform that allows for developing multi-channel experiences, generating better leads and measurable results, and scaling marketing efforts across the customer lifecycle.

3. *AdCreative.ai* generates conversion-focused ads and social media posts in seconds. It provides up to 14x higher conversion rates and is used by thousands of advertisers and brands. It offers various plans for marketers and significantly improves clickthrough rates as soon as the first month.

4. *Adext* is an advanced end-to-end marketing solution that automatically optimizes audience segments and budget allocations using AI-supported audience management automation. It maximizes revenue with proprietary machine-learning algorithms and is compatible with Google and Facebook Ads.

5. *AI-Writer* provides a platform for content creation. It uses AI algorithms to generate articles, blog posts, product descriptions, and more based on a given topic.

6. *Article Forge* automatically researches, plans, and writes long-form content. In one click, you can get a complete article of 1,500 or more words that is unique, well-written, and relevant, dramatically reducing the time and cost required to produce content.

7. *Azure Machine Learning* is a trusted platform created by Microsoft for building and managing high-quality machine learning models at scale. It accelerates time-to-value for new customers with its industry-leading matching learning operations (MLOps) and offers open-source interoperability and integrated tools. The platform prioritizes responsible AI and provides a secure environment for the machine learning lifecycle.

8. *Bardeen.ai* is an all-in-one AI automation toolkit with over 30 applications that can give a heads up for meetings, send automated emails, snap a full page screenshot and share it in one click, find emails from LinkedIn, Twitter, and GitHub, monitor websites for changes, and more.

9. *Beautiful AI* is software that automatically designs your presentations using its templates.

10. *Blaze Today* is a time management tool that helps you plan your day and increase productivity. It integrates with your calendar and to-do list to prioritize tasks and keep you on track.

11. *Brand24* helps organizations monitor their online brand reputation. It tracks mentions of a brand online and provides valuable insights about its online presence, allowing quick responses to negative comments and improving its online reputation.

12. *Brandmark.io* is a website design and branding tool that creates and manages brand identity. It provides logo design,

color palette creation, and typography selection to help brands build a cohesive and memorable brand image. Businesses use Brandmark.io to develop their brand and create a consistent look and feel across all of their marketing materials, from websites and social media to business cards and brochures.

13. *Browse AI* allows users to extract and monitor data from any website with no coding required, using robots that can be set up in two minutes. Features include data extraction, monitoring, pagination and scroll handling, scheduling, and flexible pricing.

14. *Casetext* is an AI-powered legal research platform that provides quick and accurate answers to legal questions. It has helped small law firms to work more efficiently and improve their results, driving growth and increasing their client base.

15. *Chatfuel* is a chatbot builder that allows you to create custom virtual assistants for your website, messenger, or app, all without coding. It provides a visual interface to build chatbots and has a drag-and-drop feature for easy setup.

16. *ChatGPT* is a chatbot that uses advanced language processing to understand and respond to customer inquiries. It can be integrated with websites, messengers, and other platforms to provide 24/7 customer support. It can write articles, and sales copy, alphabetize lists, provide sales ideas, and look up almost anything you ask. Think of it like Google but times 1,000! It can curate from multiple sources and answer in various language levels, from elementary to scholarly expert.

17. *ClipMaker.ai* converts YouTube videos into shortened clips to use on other social media platforms, including TikTok and Instagram. It also features templates, auto-scheduling to help users grow their accounts quickly and easily, and subtitles to increase engagement.

18. *CognitiveScale* is an AI Engineering platform that offers tools to accelerate AI deployment, enable observability and governance, and optimize business goals. It also detects, scores, and remediates AI risks.

19. *Content at Scale* is an AI content checker that uses a GPT AI detector to identify AI-generated content within seconds. It's highly accurate and trained on billions of pages of information to distinguish between humanly-optimized and AI-generated content.

20. *Conversica* provides AI-driven lead engagement software and Intelligent Virtual Assistants for marketing and sales. Their Digital Assistants conduct personalized conversations via email, SMS, and chat in the contact's language of choice. Pre-built and trained with over 1,000 conversations optimized for speed-to-market, they are a leading provider of conversation automation solutions.

21. *Copy.ai* generates text, such as product descriptions and marketing copy, using artificial intelligence. Organizations and individuals use Copy.ai to save time and effort when creating marketing or product materials.

22. *Crowdstrike* is a cybersecurity tool that uses AI to protect your devices, networks, and cloud from cyber threats. It uses machine learning and behavioral analysis to detect and prevent cyber attacks up to the minute.

23. *DALL-E* is a system created by OpenAI that generates unique and creative images from textual descriptions with AI. Essentially, you can describe an image you want, and DALL·E will generate a unique picture based on your description.

24. *DeepL Pro* is a customizable, secure, and accurate translation service that uses AI and neural networks for documents, web pages, emails, and verbal conversations. It has data security

measures and can integrate with other applications for professionals and developers.

25. *Demandbase* offers account-based marketing, advertising, sales intelligence, and data solutions for B2B companies. Its go-to-market solutions provide more innovative account intelligence through a refined view of accounts pulled from first- and third-party information, moving past spam to relevance.

26. *Design Beast* is a design tool that helps create stunning visual content in minutes. It has a library of templates, images, and design elements that you can use to create logos, presentations, and other visual content.

27. *Design Evo* is a tool that helps you create stunning graphics and logos in minutes. It uses AI and natural language processing to generate design ideas and make suggestions based on your input.

28. *Descript* is an audio and video editing tool that helps you easily edit and enhance audio and video content. It uses machine learning and natural language processing to transcribe and align audio and video files and provides a visual interface to edit the content.

29. *Dialogflow* is a Google-owned platform that builds conversational AI chatbots and other conversational interfaces. Businesses use it to automate customer support with chat agents, virtual assistants, and forms and to handle transactions through a conversational interface.

30. *Diffusionbee* uses Stable Diffusion to create and edit AI-generated art offline.

31. *DoNotPay* hails itself as "the world's first robot lawyer," offering a range of legal services and self-help tools for users,

including fighting corporations, bureaucracy, lawsuits, wage protection, finding hidden money, and more.

32. *Emplifi* is a platform that closes the customer experience gap by empathizing with customer needs and amplifying brand experiences. It offers tools to manage social media, drive revenue growth through social commerce, and create exceptional customer experiences with digital tools for the contact center.

33. *Fathom* is an AI Meeting assistant tool that can record, transcribe, highlight, and summarize meetings that can be sent to your CRM, Google Docs, Gmail, or Task manager with one click.

34. *Figma* is a design tool that allows teams to create, edit, prototype, and collaborate on design projects all in one place. It's an online platform that can be used by designers, product managers, marketers, and other creative professionals to bring their ideas to life.

35. *Finalle.ai* is a financial intelligence platform that gathers and analyzes large amounts of financial market data. It gives a comprehensive and up-to-date overview of the market and its driving forces in real-time.

36. *Fireflies.ai* is an automated meeting note-taking tool that records, transcribes, and analyzes voice conversations. It can invite Notetakers, capture video and audio, generate transcripts, create tasks, and track key metrics. Fireflies.ai integrates with several apps, such as video-conferencing apps, CRMs, and collaboration apps.

37. *Flexclip* is a video creation tool that helps you create professional-looking videos in minutes. It has a library of templates, images, and video elements that you can use to create promotional, explainer, and other types of videos.

38. *Fundwriter.ai* is known as the "Nonprofit writing assistant" and assists with writing more persuasive appeals to prospective donors that are sure to touch hearts. It helps to craft a proposal that stands out and allows nonprofits to express their gratitude to donors in no time so they can focus on what's important - making a difference.

39. *Genei* is an online research tool that can extract information from web pages and PDFs. It aims to make researching and note-taking more efficient, accurate, and hassle-free. By using Genei, you can quickly gather information and insights from any source without fearing missing important information. It is considered the best tool for note-taking and research.

40. *Genius Sheets* provides instant data analysis and allows users to generate reports and financial models within Excel or Google Sheets. It enables live data connections, automation, and connectivity with QuickBooks Online.

41. *Glasp* is a Chrome extension for writers, readers, and thinkers to easily highlight web content, save it for later and share it. It automatically curates the highlights to your Glasp homepage, which can be tagged, searched for, and shared on other platforms like Twitter, Microsoft Teams, and Slack.

42. *Grammarly* is a writing assistant that provides live feedback on grammar, spelling, and style. It improves writing communication for both professional and personal use.

43. *GrowthBar* assists teams with planning, writing and optimizing long-form content in a shorter time frame. It provides AI content generation with outlines for blogs and websites, as well as SEO with keywords, word counts, and headers. It features a 2-minute blog builder that allows users to generate a 1,500-word blog and does keyword and competitor analysis for content planning.

44. *IBM Watson Studio* uses AI to create intelligent virtual agents that provide consistent and competent customer support across all channels. It can help lower costs and enhance performance. With IBM Watson Studio, businesses can make sense of complex information, systems, and staffing issues, leading to a more efficient and productive organization.

45. *Invideo* is a video creation tool for individuals and businesses to make engaging and high-quality videos quickly and easily. It features video templates, image and text input, editing, special effects, an intuitive interface requiring no technical skills, and integration with YouTube, Vimeo, and Facebook.

46. *Jasper* is an AI platform designed for teams that want brand-specific content such as SEO content, Facebook ads, and web content in a significantly shorter amount of time.

47. *Kaiber* turns your ideas into works of art in just three clicks with its amazing video creation tool. You can use your own pictures or describe what you want, and its state-of-the-art engine will turn it into a beautiful visual story masterpiece.

48. *Kinaxis* is a supply chain management platform that uses AI to provide on-the-spot visibility, risk management, and collaboration across the entire supply chain.

49. *KNIME* creates software that simplifies data science. It offers an easy-to-use visual interface to help you analyze information without any coding. It also lets you blend, access, and visualize data. And you can integrate your favorite tools and libraries. KNIME also allows you to streamline workflow testing and validation, saving time by catching errors early.

50. *LeadSquared* is an all-in-one marketing automation and CRM software that enables lead capture, management, sales, and analytics. It streamlines sales and reduces turnaround times, allowing salespeople to work smarter and spend more time

with the right prospects. It also supports self-serve and assisted onboarding journeys for customers and partners.

51. *LeiaPix* helps graphic artists, painters, web designers, video editors, animators, and other creatives turn any image into a 3D image. It provides an easy-to-use interface, allowing users to quickly and easily transform photos into 3D without requiring any special technical skills.

52. *Legalese Decoder* translates legal language or "legalese" into plain, easy-to-understand language. This tool aims to make legal documents, contracts, terms and conditions, and other legal agreements more accessible and understandable to the general public.

53. *Lensa* is a technology company that provides job search and recruitment services to businesses. Lensa uses artificial intelligence to match job seekers with job openings based on factors such as skills, experience, and qualifications. Companies use Lensa to find suitable candidates for their job openings more efficiently and effectively.

54. *Linguix* is an AI tool that provides writing assistance. It helps users improve their writing skills by giving grammar, style, tone, and more suggestions as they write.

55. *Looka* is a logo design platform that allows users to generate and customize their logo in minutes. It can also refine existing designs and allows you to purchase various versions and sizes of your logo for different uses. Looka requires no design skills or experience to create a brand identity.

56. *Marketing Blocks.ai* is an all-in-one AI marketing assistant designed to produce landing pages, promo videos, ads, marketing copy, graphics, email swipes, voiceovers, blog posts, articles, and art in minutes. With just a keyword, AI creates

a profitable online business that can be used for your own business or sold to clients.

57. *Melville* is a podcast copywriter that transcribes podcast episodes and writes show notes for them. The output will need some editing, but Melville does most of the work. There are no limits on the number of podcasts you can have in your account, and you can upload files in MP3 format.

58. *Memorable Admaker* is a tool that generates images optimized for marketing KPIs, trained on human reactions to content for saliency and memorability. The tool helps pretest and optimize creative strategies with high accuracy, validated by human behavior testing.

59. *Merlin* makes it a breeze to summarize content from any website, automatically create formulas or code with your guidance, write professional email responses, condense long documents, and develop new ideas for your marketing efforts.

60. *Midjourney* is an AI image generation bot that creates unique images from textual descriptions, commonly known as prompts, input by the user.

61. *Mindsmith* is a learning platform that simplifies course authoring, allowing you to create engaging lessons easily. Its AI Lesson Assistant generates high-quality content, including entire lessons and training modules, while providing customizable question insertion for more digestible learning.

62. *Mixo* is a website builder that helps entrepreneurs quickly launch and validate their business ideas with a brief idea description. It generates website content, landing pages, and a built-in email waiting list, allowing pre-launching products, gathering insights, building waiting lists, and running beta testing programs.

63. **Mobilemonkey** is an AI tool that provides chatbot services. It helps companies create and manage AI virtual assistants for customer service, sales, and marketing, without the need for any coding skills.

64. **Movio** is an AI video generator that quickly creates professional-quality videos for marketing, sales, training, and learning. It offers customizable avatars with several accents in different languages, background music, and 200+ templates, allowing users to create videos in minutes with no editing skills required.

65. **Newswriter.ai** is a writing tool that creates compelling press releases in minutes using GPT-3 OpenAI technology. It offers services to write a press release from scratch or improve an existing one. It also provides news marketing tools to distribute news to Google News and hundreds of other websites.

66. **OpenCV** helps organizations analyze and understand visual information, such as images and videos, to extract useful information. It can be used for various tasks, including face recognition, tracking objects, and analyzing traffic patterns. A company might use OpenCV to build a security system to detect when someone enters a restricted area. OpenCV also offers AI courses on Deep Learning, Computer Vision, and related subjects and provides consulting services led by its team of experts.

67. **Optimity** has demand planning software that creates accurate demand plans to prepare for volatility. The software uses statistical forecasting techniques, collaboration tools, and analytics to help ensure supply chain performance and profitability.

68. **Optimove** improves customer relationship management (CRM) strategies by providing actionable insights and

personalized recommendations for better customer engagement, retention, and loyalty.

69. *Oracle Eloquais* is a B2B marketing automation platform that assists marketers in managing campaigns, lead generation, and customer understanding. It features AI-enabled tools and multi-channel campaign creation capabilities. The platform aims to increase conversions and sales by targeting customers across various marketing channels.

70. *Otter* provides instant transcription and note-taking services. It transcribes audio and video instantaneously, making it easy to take notes, save highlights, and access important information later.

71. *Persado* uses advanced deep learning models and a decision engine to generate language that motivates individuals to engage and act. Their language outperforms the best copy 96% of the time. It drives unprecedented conversion rates and revenue growth for leading brands, leveraging a vast marketing knowledge base and learning consumer response patterns to deliver hyper-personalization at scale.

72. *Photor AI* is an image recognition tool that analyzes photos for professional or personal use online. It selects the best images using machine learning technology and technical aspects such as brightness, contrast, and noise. Users can easily upload multiple photos and choose the best ones.

73. *Phrasee* increases customer clicks, conversions, and buyers by delivering optimized marketing messages at scale. It powers billions of marketing messages across different channels such as email, push notifications, SMS, social media, and the web. The platform ensures that the messages are on-brand and designed to perform at a high level.

74. *Play.ht* is a voice generator and realistic text-to-speech (TTS) technology that instantly converts text into natural-sounding speech in various languages and accents. Users can download the audio files. The platform provides a growing library of over 900 AI-generated voices with human-like intonation powered by machine learning technology.

75. *Podcastle* is a podcast hosting and publishing service. It makes it easy for anyone to start a podcast, regardless of their technical expertise. With Podcastle, you can upload and host your audio files, create show notes, and distribute your podcast to popular platforms like Apple Podcasts, Spotify, and more.

76. *ProwritingAid* provides writing assistance to help users improve their writing skills by offering suggestions on grammar, style, tone, and more as they write.

77. *Quillbot* is a writing tool that helps users rephrase and improve their writing. It uses advanced algorithms to analyze writing and provide suggestions for improvement.

78. *ReachOut.AI* is a video prospecting platform that helps entrepreneurs and sales teams generate personalized videos, increase email response rates, and access a sales engagement CRM. It offers video creation with AI-generated spokespeople with dynamic backgrounds.

79. *RELEX Solutions* is a supply chain management platform that provides demand forecasting, inventory optimization, and replenishment planning. It produces fast information through real-time data processing, allowing planners to concentrate on innovation while automating repetitive tasks.

80. *Replit* is an online platform for coding and programming. It provides a simple and accessible environment for people of all skill levels to write, run, and share their code with others. You can write code in many different programming

languages and easily run it to see the results, all in one place. Replit aims to make coding more accessible and collaborative so anyone can learn and create with code.

81. *Runway* offers a comprehensive content creation suite that allows you to effortlessly remove video backgrounds and objects, edit videos concurrently with your team, and much more. With over 30 cutting-edge AI tools, Runway is revolutionizing how you create content, making it easier and more efficient than ever.

82. *Ryter* is a platform that creates and publishes web pages, blogs, and other digital content. You can easily create a website or blog without having technical skills or coding knowledge.

83. *Salesforce Einstein* is a comprehensive AI technology for CRM that enables businesses to anticipate opportunities, resolve issues before they happen, personalize experiences, and create more innovative apps for employees and customers. Einstein helps sales, service, marketing, and IT teams to be more predictive and proactive in meeting customer needs.

84. *Salesforce Pardot* is a B2B marketing automation solution that helps organizations create and manage marketing campaigns and track and measure their effectiveness. It offers a 360-degree view of prospects and supports various marketing channels.

85. *Sapling* is a language model copilot that assists customer-facing teams with responding more quickly to customer inquiries. It provides real-time suggestions on messaging platforms and CRMs, enabling sales, support, and success teams to compose personalized responses more efficiently.

86. *Scalenut* is an SEO and content marketing platform that increases organic online traffic. It is ideal for SEO strategists, marketers, agencies, and founders to discover and create the

most relevant content for their customers. Scalenut uses deep learning and AI to produce high-quality content.

87. **SEO.ai** is an AI platform that assists SEO marketers in quickly creating high-quality content that performs well, with features like keyword research, content suggestions, competitor analysis, and AI-assisted copywriting in 50 languages.

88. **SharpSpring** by Constant Contact offers a full-funnel Marketing & Sales Automation + CRM solution for agencies and SMBs. Its features include email marketing, marketing automation, sales engagement, CRM, social media, retargeting ads, and tracking & analytics. The platform helps organizations reach and engage their audience, equip teams to close deals faster, capture every opportunity, and convert more prospects.

89. **SidekickAI** is a software that simplifies scheduling with AI and language processing. It analyzes availability and suggests alternative dates and times when necessary. It can instantly book meetings, send confirmations, add them to calendars, or email guests to complete the meeting request.

90. **Simplified** is a design and collaboration platform with one-click graphic design tools, AI copywriting, animation, and video and audio editing tools. It offers millions of free photos, thousands of templates, a content rewriter, social media management, and more to help users easily create, edit, and publish content.

91. **Social Media GPT** is a chrome extension that uses ChatGPT to generate engaging comments on various social media platforms automatically.

92. **Soundraws** is an audio production platform that makes producing high-quality audio content easier for content creators. It uses machine learning algorithms to automatically generate

audio tracks based on user-defined parameters like genre, tempo, and style. Soundraws make audio production faster, more efficient, and more accessible.

93. *Synthesia.io* is an AI video creation platform that allows you to create videos in minutes from plain text. It offers digital twins of real actors as avatars and digital clones of real people's voices as AI voices, enabling users to create videos in 120 languages without being on camera or recording their voices.

94. *Tailor Brands* is a branding solution suite of tools that creates custom designs and logos. With the help of machine learning algorithms, Tailor Brands suggests unique designs and logos based on the input given by the user.

95. *TensorFlow* is a program that helps businesses build and train smart computer systems. It uses a type of AI called machine learning to help computers learn. Its AI technology solves various problems, like recognizing objects in images, translating languages, and predicting what will happen in the future based on past events.

96. *Tome* is a platform that uses AI technology to help people create better stories and presentations. With its AI-enabled storytelling format, you can type in a prompt and watch as Tome generates entire narratives from scratch or adds extra content in just seconds. Using advanced AI tools like GPT-3 and DALL-E 2, Tome makes it easy to create great presentations with minimal effort.

97. *Toonly* uses AI to create animated explainer videos. It offers a user-friendly interface to help users easily create animations, even if they don't have any prior experience.

98. *UiPath* is an AI-powered platform that automates monotonous and manual supply chain tasks, allowing professionals

more time to prioritize higher-value tasks. UiPath automates workflows, improves efficiency, and provides insights for cost-effective digital transformation while minimizing disruption.

99. *Vanta* is a platform that automates the compliance process for companies for security and privacy standards like SOC 2, ISO 27001, HIPAA, and GDPR. It does this by continuously monitoring and improving a company's security practices, making it easier to pass security audits. Vanta simplifies security procedures and gets companies audit-ready quickly without spending months preparing.

100. *Voicemaker* is a text-to-voice platform that generates custom voiceovers for videos, presentations, podcasts, and other media. It provides a range of voices to choose from and allows users to input their own text to generate a voiceover.

Whether you desire to increase your visibility, customer base, sales revenue, or all three, there is an AI tool that is perfect for your business needs. With these tools, you can streamline your workflow, create custom AI solutions, experiment with the latest AI models, and discover creative ways to implement these platforms to improve your company's operations. Have fun taking these tools for a spin!

| 9 |

Building an AI Dream Team

> *"Harnessing machine learning can be transformational, but for it to be successful, enterprises need leadership from the top. This means understanding that when machine learning changes one part of the business – the product mix, for example – then other parts must also change. This can include everything from marketing and production to supply chain, and even hiring and incentive systems."* – **Erik Brynjolfsson, Director of the MIT initiative on the digital economy**

Knowing what it takes to assemble a group of individuals with the necessary skills will help companies implement the right AI strategy successfully. We will delve into the importance of hiring the right people, fostering a positive and productive work environment, and creating a culture that encourages innovation and creativity. Whether you're an entrepreneur, a manager of a team, or running a large company, these next steps are for you. You'll learn about the benefits of building a strong AI team and the proper steps to create your dream team.

The importance of having a rockstar AI team

Having the right team in place is essential for any company looking to grow and thrive in the age of artificial intelligence. Having the right talent on your team is the key to unlocking its full potential. When building an AI team, looking for individuals with a combination of technical skills, business acumen, and creativity is essential. You'll want team members who understand the latest AI technologies and platforms and how to apply those technologies to your business. An understanding of your industry is helpful, but as long as your AI team can achieve the desired results, that will still get you where you want to go.

Depending on the complexity of your business, you may need a data scientist on your team who is knowledgeable about machine learning algorithms to help you make informed decisions and optimize your workflows. Software engineers are also a helpful addition to an AI team in developing and implementing AI tools, such as chatbots or recommendation engines, to improve the customer experience and increase sales. And if you are a small agency, don't worry. It is common for smaller and less complex companies to train their current in-house talent on new AI platforms or outsource the necessary skill sets. Suppose you're a solopreneur or independent consultant. In that case, you can easily learn many of the "ready-to-use" AI platforms in your business by watching YouTube tutorials or enlisting the expertise of the AI software companies you use.

In addition to having the right technical skills, it's also essential to have a well-rounded and collaborative team. You'll want team members who can collaborate effectively, share ideas, and communicate clearly to ensure your AI initiatives succeed. It's also essential to have team members who are innovative, as this quality will help you stay ahead of the curve and continuously improve your AI capabilities. Having a diverse team with different backgrounds and perspectives is also essential. This can help you to understand your customers better, create inclusive products and services, and avoid partiality in your AI systems. Additionally, a diverse team can bring new ideas and approaches to problem-solving, which can help you stay ahead of the competition.

Creating a culture that supports and fosters your AI team is important. Working with a team that does not embrace technology or that resists change is incredibly difficult. Every successful team should at least have the right tools and resources along with continuous learning and development for the ever-evolving AI technology. By creating this supportive work environment, you can attract and retain top talent and ensure your AI initiatives succeed.

AI is making many companies reconsider their current business methods and examine whether there is an AI tool that can make them more efficient. Companies that have embraced AI have seen a significant improvement in their operations and have gained a competitive edge over those that have not. Having the right AI team in place is essential for any company looking to scale in the smartest and fastest ways possible. By assembling a talented and dedicated team with a combination of technical skills and creative thinking and fostering a supportive work environment, you can unlock the full potential of AI and use it to scale your company exponentially.

When considering working with tech pros to improve your company's workflows with AI, here are some critical questions to ask. Asking these questions is a great exercise to work through with key people in your organization.

- What are our goals for using AI?
- What problems do we currently have that AI might be able to fix?
- How do we want AI to help us reach these goals?
- What's the current state of our AI setup?

- What areas of our AI need improvement?
- What's the best way to use AI in our industry?
- Which AI tech is best for our needs?
- What challenges and limitations can we anticipate with AI?
- How can we keep our data safe and secure in our AI processes?
- What's the plan for implementing AI, including time and budget?
- How will we measure the success of our AI implementation?

Asking these questions will help you get a clearer picture of your goals and objectives, understand your current AI situation, identify areas for improvement, manage reasonable expectations, and lastly, make sure you have a smooth and successful implementation of AI in your workflow.

How to build an AI team that gets results

Building an AI team is an important step in any organization's journey to harness the power of artificial intelligence. It's essential to start with a clear understanding of what the team will be responsible for and what skills and experience are required to perform the job effectively.

Defining the Role: The first step in building an AI team is to determine the role and responsibilities of each member. Defined roles will help you understand what your AI team will be working on and what skills and experience you need to look for when hiring. Some regular responsibilities for AI teams include automating tasks, improving the customer experience, analyzing information, and overseeing implementation.

Writing Job Descriptions: Once you have a clear understanding of the role, it's time to write accurate job descriptions that clearly outline the responsibilities, qualifications, and requirements for each position. Make sure to include a detailed list of the tasks the AI team will be responsible for and the necessary skills and experience required to perform those tasks. Some common skill sets needed for building a more complex AI solution are programming languages (i.e. Python, R, Java), machine learning, deep learning, and data engineering. If tapping in-house talent or outsourcing talent are not available options, then finding candidates to hire is the next logical step.

Finding Candidates: The next step is finding potential AI candidates for the team. There are several ways to do this, such as posting job listings on websites like Indeed or LinkedIn. You can also use referrals from current employees, attend job fairs, or work with a tech recruitment agency.

Recruitment and Screening: Once you have a pool of potential candidates, it's time to start the recruitment and screening process. This process typically involves reviewing resumes, conducting initial phone interviews, and inviting the most qualified candidates for in-person or virtual interviews. The goal is to find the best candidates who meet the requirements for the role and will fit well within the team and the organization.

Interviewing and Selecting: During the interview process, ask questions that will help you understand the candidate's experience, qualifications, and ability to perform the tasks required for the role. After the interviews, then make a final selection and offer the job to the chosen candidate.

Onboarding and Training: Finally, onboard and train the new AI team members. Onboarding involves introducing them to the company culture, policies, and procedures, as well as providing any necessary training to help them perform their role effectively. The onboarding process should be thorough to ensure the new team members can hit the ground running and make a positive impact as soon as possible.

Finally, building an AI team requires careful planning and consideration. Companies need an AI team to navigate the complex and rapidly changing technological landscape, take advantage of the potential benefits of AI technologies, and overcome the challenges associated with implementation. With a highly skilled and diverse team in place, businesses can position themselves to succeed in the digital age and stay ahead of the competition. By seeing this process through from beginning to end, you can build a strong AI team that will drive your organization forward.

| 10 |

AI Ethics and Best Practices - Doing it Right

> *"I'm increasingly inclined to think that there should be some regulatory oversight, maybe at the national and international level, just to make sure that we don't do something very foolish."* – **Elon Musk, CEO of Tesla, SpaceX and co-founder of OpenAI**

As we delve deeper into the world of AI, consider not only its potential benefits but also its ethical implications. With the increasing use of AI in various industries, it's crucial to establish best practices and guidelines to ensure it's being used responsibly and ethically. Learning about data privacy, algorithmic bias, and accountability is necessary so you can be confident that you're using AI in a way that aligns with your company's values and benefits society as a whole. Whether you're running a company or an influencer over technology on your job, this chapter is essential reading to help you do AI right.

Ethical considerations for AI and why it's important

The use of artificial intelligence in companies has become increasingly prevalent in recent years, and with this growing popularity comes the need for ethical considerations. The ethical use of AI is critical not just for maintaining the reputation of companies but also for ensuring that AI is used in ways that are not harmful to others.

One of the primary ethical considerations of AI in companies is the issue of data privacy and security. AI algorithms rely on large quantities of information to learn, and companies must be mindful of the data they collect and how it is used. Proper data use includes ensuring that information is collected and handled according to privacy laws and that appropriate measures are in place to protect sensitive information.

Another important ethical consideration is the potential for AI algorithms to perpetuate unfairness and discrimination. AI algorithms are trained on data; if biased or discriminatory, the output will also reflect these distortions in how it processes information. Companies need to be mindful of the data they use to train AI algorithms and work to address any potential partiality within it.

In addition to these concerns, there is also the question of accountability when it comes to using AI. If an AI algorithm produces an outcome that has a negative impact on an individual or group, it can be challenging to determine who is responsible for this outcome. Companies need to be transparent about their AI algorithms and ensure that appropriate measures are in place to hold individuals and organizations accountable for any adverse results.

The ethical use of AI is also important because it builds trust with customers and stakeholders. Companies that use AI ethically and responsibly are more likely to be trusted by their customers and stakeholders. This trust can help build long-term relationships, increase support for their products and services, and avoid costly legal trouble.

Another ethical consideration in using AI is the potential impact on employment and the workforce. As AI algorithms become more sophisticated, there is a risk that they may replace human workers, leading to job losses and economic inequality. Companies need to be aware of the potential impact of AI on the workforce and work to ensure that the benefits of AI are shared in a way that supports economic growth and stability.

To ensure the ethical use of AI, companies must adopt a proactive approach beyond simply complying with laws and regulations. Developing a set of ethical principles and guidelines for using AI embeds these principles into the company's culture and core values. One way to achieve this includes creating an ethical review board responsible for overseeing the use of AI and ensuring that it aligns with the company's ethical principles.

A critical step in ensuring the ethical use of AI is to involve stakeholders. Stakeholder inclusion can consist of engaging with customers, employees, and other stakeholders to understand their perspectives and ensure that any concerns about AI use are considered. Stakeholder engagement programs, public consultations, and the creation of user panels can achieve the transparency companies need to remain accountable to those they serve.

Finally, it is necessary for companies to be transparent about the use of AI and to communicate openly and honestly with stakeholders about the benefits and limitations of AI. Transparency can build trust with stakeholders and ensure that AI is used properly. Building this trust is crucial in using AI because it is not always understood or perceived favorably. As AI is adopted by the mainstream, it is inevitable that it will cause controversy, even to the extent that some believe that AI will doom humanity.

The ethical use of AI in companies will help put customers' minds at ease. Keep in mind that AI is relatively new to many people, and people generally fear what they do not understand. Companies need to be mindful of data privacy and security, the potential for AI algorithms to perpetuate partiality and discrimination, and the issue of accountability when it comes to using AI. It's important to remember that AI is only as good as the information you feed it. By considering these ethical considerations, companies can build trust with their customers and stakeholders and ensure that AI is used to benefit everyone.

Let's address the elephant in the room: Is AI immoral?

Unethical and immoral are often used interchangeably, but they actually mean two different things. Unethical refers to actions that go against established moral and ethical standards that may not necessarily be illegal. One such example is a company that engages in environmental pollution may be considered unethical because it goes against societal norms of preserving the environment. On the other hand, immoral refers to actions that are considered wrong or evil by nature. Lying to someone, cheating, or stealing are considered immoral actions because they are generally accepted as inherently wrong. It's important to understand these two concepts' differences to have the wherewithal to apply a moral and ethical standard while using AI and to avoid harming people.

Trust me. I've had my fair share of concerns about how far technology will go and what will happen to society when the bad guys get their hands on it. But as the saying goes, don't throw the baby out with the bath water! Anything good can be used for evil purposes, so my focus is **only** on doing good with it.

First, it's vital to understand that many people are concerned about the potential impact of AI on society and fear that it might become uncontrollable or be used for malicious purposes. So if you are one of these individuals, you are not alone and your concerns are valid. It's imperative for me to address some common concerns here in a thoughtful and informed way. Deep fakes, voice cloning, spreading propaganda through algorithms, AI-generated art that mimics real artists' work, brain implants, and more can venture into the dark side of AI.

One way to overcome the moral dilemma is to concentrate on the potential benefits of AI. AI can help us solve complex problems, improve healthcare outcomes, and increase efficiency in various industries. When we focus on these benefits, it becomes easier to see AI as a tool that can be used for good rather than evil. It's also imperative to recognize that AI, in itself, is not inherently good or evil. It's a tool, and how it's used depends on the motivations and intentions of those who use or control it. For instance, the internet can be used for education and communication, but it can also be used to spread hate and misinformation. The greater concern is the level of integrity of people who create AI tools and their users. That means that you and I have a responsibility to make sure that we use AI to benefit and not harm others. Ideally, AI would be controlled and used only for good, but we know that is not guaranteed.

Another way to overcome the moral dilemma over AI is to emphasize the growing awareness of the need to ensure that AI is designed, developed, and used ethically and responsibly. When this book was published in early 2023, no substantial legislation on AI in the United States had been passed. However, large tech companies have begun to discuss policy concerns about data and innovative technology, which will pave the way to comprehensive legislation.

One such tech company is IBM and its launch of The IBM Policy Lab. The IBM Policy Lab is a forum that provides policymakers with recommendations to embrace innovation while maintaining trust in a world being reshaped by data. They collaborate on public policies to address future challenges as organizations and governments continue deploying new technologies that positively transform the world. I predict that with increased AI adoption, legislation from regulatory authorities will emerge to define what is legal and what is not.

Another argument about whether or not AI is moral or immoral is the idea that AI systems are simply following the rules and programming set by their creators. Furthermore, AI does not have a conscious or free will and consequently cannot be considered immoral or evil in the same way human beings can.

To sum it all up, the moral dilemma surrounding AI can be overcome by focusing on the potential benefits of AI, emphasizing the importance of ethical considerations, and recognizing that AI is a tool that can be used for good or evil, depending on the user. Additionally, regular citizens like you and I can hold our lawmakers accountable for creating and passing laws that protect the public from ill-intended uses of AI. By approaching the topic with an open mind and considering all perspectives, we can work towards a future where AI is used to improve our lives in the most ethical way possible.

Are there downsides to using AI in business?

Well, for every upside, there is almost always a downside. Now that we've addressed the ethical and moral dilemmas some companies might face when using AI, let's explore its potential negative impact on businesses. Since much of what you've read so far has focused on all of the things that AI can do to help your company to scale, such as increasing efficiency and productivity, smart strategizing, and lower costs, I'd be remiss and somewhat imbalanced if

I didn't provide some real examples of the challenges that AI might present to your business. One thing to remember here is that your organization's size and complexity will determine the magnitude of the downside. The negative impacts will likely lessen with smaller operations such as microenterprises, consultants, and solopreneurs.

Job loss. One of the biggest concerns about AI is its potential to replace jobs and reduce the need for human labor. In economics, this is known as creative destruction. Job losses and increased unemployment can harm the economy and local communities. Many retail jobs are at risk of being replaced as retailers adopt self-checkout systems and other AI-supported technologies. However, just as some jobs will be eliminated, new ones requiring different skill sets will be created because of AI.

Bias and discrimination. Again, AI algorithms are only as good as the data on which they are trained. If this data is biased or discriminatory, it can lead to unfair outcomes and discrimination, damaging an organization's reputation and limiting its growth potential. For example, facial recognition technology has been criticized for being unjust against certain ethnic groups and leading to wrongful arrests.

Technical challenges. Implementing AI can be complex and challenging, and many businesses need a solid technical foundation to get the most out of the technology. This challenge can lead to delays, increased costs, and a lack of progress. Technical complexity can also lead to system failures and compatibility issues, which can impede the success of AI initiatives. To avoid technical challenges, companies can always search for AI solutions that promise a more straightforward implementation.

Privacy and Security Concerns. AI systems often process and store an extensive amount of sensitive information, which can pose significant privacy and security risks. Data breaches, theft of sensitive information, and other security incidents can negatively impact a business's reputation and bottom line. Overcoming this challenge requires having robust privacy and security policies in place and ensuring that their AI systems are adequately secured and monitored. AI systems and their data might be vulnerable to hacking and other cybercrimes. The rise in cyber-attacks highlights the need for organizations to have solid cybersecurity measures in place to protect against AI-related security risks.

Decreased human creativity and initiative. AI systems are designed to use data-based algorithms to do what humans can do but only more efficiently and effectively. This can reduce creativity and initiative among employees, as they rely on AI systems to do the work for them. For example, suppose an AI system is used to generate reports and insights. In that case, employees may become less likely to think critically and creatively about the information, decreasing the quality and value of their contributions.

Too much dependency on AI systems might also cause employees to become less likely to experiment with new ideas and approaches, which could lead to a decrease in innovative products and services and diminished competitiveness. It's also worth noting that AI is not a substitute for divine revelation or inspiration. It's important not to idolize AI technology at the risk of stifling superior creativity.

Keeping up with Increased regulation. As AI becomes more widely integrated into various industries, governments and other organizations will likely introduce new regulations to govern its use. For instance, the EU's General Data Protection Regulation (GDPR) has required businesses to implement new procedures to protect the personal information of their customers and employees. New regulations and policies can inundate companies with complex rules and added costs, which can stifle growth. AI systems will become subject to increased regulations as it evolves and becomes more widely adopted. Failing to comply with these regulations may result in significant legal and financial penalties and harm a company's reputation. To overcome this challenge, organizations must stay abreast of the laws and regulations that apply to AI and work with legal and compliance experts to ensure their AI systems are compliant.

Difficulty attracting and retaining talent. As AI becomes more widely adopted, employees with skills and expertise in AI technology will be in higher demand. The newness of this field can make it difficult to find, attract, and retain the talent they need to compete and succeed in the AI-driven economy. Difficulty finding and retaining employees with the right technical skills to build and maintain AI systems can affect the ability to scale. Getting past this hurdle requires proactively investing in training and development programs for existing employees, partnering with AI experts, or outsourcing projects to specialized AI firms.

It is essential to take a strategic approach to AI, considering both its potential benefits and challenges so that companies can leverage the technology to achieve their goals and have realistic expectations. Steps to mitigate the potential challenges and risks associated with AI may involve:

- Working with AI experts and legal professionals.
- Conducting thorough risk assessments.
- Investing in ongoing training and development for employees.
- Being transparent with customers and employees about how AI is being used.

By taking a balanced approach, businesses can harness the power of AI to move forward while also reducing potential risks and challenges. As AI continues to evolve, organizations that invest in technology and adapt to the changing landscape will be well-positioned to scale faster than their competitors.

Best practices for using AI responsibly

Conduct a thorough risk assessment. Before implementing any AI system, a company might conduct a risk assessment to identify potential legal and ethical risks. This assessment might consider the possibility of data breaches that could result in sensitive information being leaked. The company might also consider the potential for bias in the AI system's processing and how it could affect customers.

Ensure data privacy. A healthcare organization using AI to analyze patient data, for example, must ensure that all personal information is collected, stored, and processed, following relevant laws and regulations. All organizations should implement data security measures to prevent unauthorized access to sensitive information to protect customers' privacy rights.

Avoid bias in data. An AI system used in the hiring process might perpetuate existing prejudices if the data used to train it isn't very objective. To avoid this, the company might ensure that the data used to train the AI system is impartial and regularly audited to identify and address any partialities that may arise.

Be transparent about AI use. A financial services company using AI to approve or deny loans should be transparent about how its AI system uses data to inform customers about the factors that influence outcomes. A company might provide customers with a report that explains how their credit score, employment history, and other factors influenced the AI system's results.

Establish clear governance and accountability. An organization using AI to detect fraud, for example, should establish clear governance and accountability frameworks to make sure that the technology is used properly. The organization should have clear policies for using AI, procedures for managing and mitigating legal and ethical risks, and regular audits of the technology's performance.

These are just some best practices for using AI to avoid legal and ethical pitfalls. By following industry and legal guidelines, organizations can ensure that their use of AI is responsible, ethical, and compliant with relevant laws and regulations.

Other common challenges with AI adoption

Adopting AI can bring significant benefits, but it also comes with its own set of challenges. Understanding and overcoming these challenges can be critical to the success of AI initiatives and getting the most out of them. This section will discuss some common challenges of adopting AI and strategies for overcoming them. The more complex your operations are, the more complex the adoption process.

Data Quality and Availability. One of the greatest challenges of adopting AI is having the correct data to train and operate the AI models. Extracting and sifting through data can be complex and time-consuming, especially if this information is stored in different locations or formats. A comprehensive data management strategy that includes quality checks, governance policies, and integration must be developed to overcome this challenge.

Integration with Existing Systems. Integrating AI systems with existing workflows and software programs can be complex and time-consuming, resulting in compatibility issues, system failures, and frustrated employees. To tackle this issue, enlisting the assistance of AI experts will help ensure that systems are properly integrated and that there is a clear plan for how AI will be integrated with existing systems over time. Companies can also rely on the expertise of in-house talent or outsourced AI technology firms to make sure that their systems are correctly configured or choose AI solutions specifically designed for ease of use and implementation.

Lack of Understanding and Acceptance. A lack of understanding and acceptance among employees, customers, and other stakeholders can present a challenge. Resistance to change can make getting buy-in for AI initiatives challenging and lead to resistance and pushback. To resolve this challenge, transparency about AI initiatives and engaging employees, customers, and other stakeholders in discussions about the benefits and risks of AI is crucial. Furthermore, implementing gradual changes will also lessen resistance and help prevent abrupt changes.

Costs. Implementing AI systems can sometimes be expensive, especially for large, complex organizations, and this may require a significant upfront investment in hardware, software, and personnel. Businesses should carefully consider AI's costs and return on investment and explore cost-effective solutions such as cloud-based AI and third-party services to reduce expenses.

Measuring the Impact and ROI. Finally, a common challenge of adopting AI is measuring their AI initiatives' impact and return on investment (ROI). It can be difficult because AI systems can have indirect and hard-to-measure benefits, such as improved customer engagement or increased efficiency. Developing a clear framework for measuring the impact of AI and tracking key metrics, such as customer satisfaction and operational efficiency, over time will help navigate this issue.

The challenges faced by businesses when adopting AI are varied and complex. By addressing these challenges, organizations can overcome the hurdles, realize the full potential of the technology, ensure the success of their AI initiatives, and avoid many of the pitfalls already mentioned. When implemented strategically, the AI adoption process does not have to be daunting or intimidating. Know the limitations of your team and outsource help when necessary. When done right, the risks will outweigh the rewards in the end!

Incorporating technology governance and security measures

AI governance refers to the policies, procedures, and practices that organizations implement to ensure the responsible use of AI and minimize the risks associated with its use. Here's a look at what some tech giants are currently doing to incorporate AI governance within their policies and procedures.

Google has established principles to ensure that their development of AI is responsible and will not pursue AI applications in certain areas, including those that cause overall harm, facilitate injury to people, violate privacy norms, and contravene international law and human rights. Their six objectives for AI applications are: socially beneficial, avoiding bias, ensuring safety, being accountable to people, incorporating privacy principles, upholding scientific excellence, and making them available for uses that align with their principles. The company acknowledges that the field of AI is constantly evolving and will approach its work with humility, engage with internal and external stakeholders, and adapt its approach over time (ai.google, 2023).

IBM has identified five foundational pillars of trustworthy AI: Explainability, Fairness, Robustness, Transparency, and Privacy. These pillars are crucial for the development, deployment, and use of its AI systems. To employ these ethical practices, IBM is committed to evolving these practices as AI capabilities increase. The company has developed five practices of everyday ethics to build and use AI systems alongside the five pillars of trustworthy AI. These practices include taking accountability for AI outcomes, being sensitive to cultural norms, addressing biases, promoting inclusivity for human understanding of AI decision-making, and preserving users' control over their data. By following these practices, IBM intends to create more ethical and trustworthy AI systems Cutler & Pribić, 2022).

Microsoft's building blocks of its responsible AI program are designed to reflect the principles it has adopted. These principles include a governance structure, rules to standardize its responsible AI requirements, training for employees, and tools and processes for implementation. Its governance approach follows the hub-and-spoke model that integrates privacy, security, and accessibility into its products and services. The "hub" includes the Aether Committee, the Office of Responsible AI, and the Responsible AI Strategy in Engineering (RAISE) group, while the "spokes" include the Responsible AI Champs community, appointed by Microsoft leadership and located across the firm. In addition, the company has created a process for ongoing review and oversight of high-impact cases (Crampton, 2021).

Amazon strives to be customer-centric, using machine learning to constantly improve its products and services. To protect customer data, they use privacy-enhancing technologies and have company-wide policies that govern how data is processed and stored. One area of innovation is differential privacy, which limits the amount of information about individuals that can be recovered from the output of a data analysis algorithm. Differential privacy ensures algorithms can learn frequent patterns in data without memorizing details about any specific individual. Amazon has explored new methods to achieve the same privacy with better accuracy and continues to evolve in its AI governance and best practices (Amazon, 2018).

Accenture has developed a Responsible AI Framework that consists of four key pillars. The first pillar involves refining capabilities into guiding principles and establishing governance structures. The second pillar deals with managing risks, policies, and controls to comply with data and AI ethics regulations. The third pillar provides tools and techniques to support the implementation of refined capabilities. Finally, the fourth pillar involves educating employees to enable them to adopt refined capabilities into their day-to-day operations. Accenture considers the risks that AI poses to safety, privacy, and human rights and the magnitude of the potentially impacted audience. As such, it strives to take the appropriate course of action for mitigating any existing or potential risks (Daugherty, 2021).

More and more companies of all sizes will be faced with incorporating AI governance into their policies and procedures as AI technology advances and becomes more widely used. By establishing ethical principles, reviewing and approving AI projects, and providing employee training, companies can be more confident that their use of AI is responsible, ethical, compliant with relevant laws and regulations, and in line with their values. As AI continues to become an increasingly important technology, it's encouraging to see that companies are taking the necessary steps to ensure that AI is used in used appropriately.

| 11 |

The Future of AI - Ready, Set, Go!

> *"It's going to be interesting to see how society deals with artificial intelligence, but it will definitely be cool."* **— Colin Angle, co-founder of iRobot**

Are you ready to see what the future holds for AI? We'll take a closer look at where AI is headed and what it means for the future. From advancements in machine learning to the integration of AI into various industries, we'll explore the exciting possibilities that lie ahead.

Predictions for the future of AI and what's next

At the time of the release of this book, it's 2023, and AI has come a long way since its inception and will continue to evolve rapidly. We now have AI-powered robots that can do the dishes, flip hamburgers, wait tables, help track inventory, and perform certain surgeries. AI personal assistants can schedule our day, and AI cars can drive us around town. This all sounds eerily similar to the 1980s cartoon series that I watched as a kid, *The Jetsons!* It all seemed so far-fetched then, and who would have thought that eight of the advanced technologies that appeared in the cartoon series would actually exist today? As a matter of fact, jetpacks, holograms, drones, smartwatches, flying cars, robot housekeepers, 3D printed food, and smart shoes, have all existed since as early as 2015 (Zipkin 2015)! But what's next for robots and AI technology overall?

I believe AI will continue to blow our minds because its future is going to take us all on a rollercoaster ride. First, AI will be programmed to become sophisticated to handle more human-like tasks and in its ability to recognize and understand emotions. Imagine having a conversation with your AI personal assistant, and it can tell when you're happy, sad, or angry. Another exciting development is the integration of AI into our daily lives. In the future, more smart homes that are AI-optimized homes that control our lights, heating, energy usage, and even coffee machines will become the standard. And if that wasn't enough, we'll also have AI-enabled fashion, where our clothing can change color and style based on our mood and surroundings.

AI is also set to revolutionize the healthcare industry, with AI-based diagnoses, treatments, and surgeries becoming common. AI-enabled wearable devices will become more advanced, allowing us to monitor more details about our health in real-time and get personalized health advice based on our lifestyle and biometrics. There will also be more tailored health and wellness programs at companies. AI will become your personal health and wellness guru, using data to give you even more detailed tips and tricks to reach your health goals, like a personal health coach. And the best part? It will help patients become more healthy while reducing healthcare companies' operational expenses.

Now, this is a fascinating topic: Robots as surgeons! These aren't your average robots, but they are high-tech machines that can precisely perform complex procedures, making surgeries safer and more effective than ever. Think of it as having a highly skilled surgeon inside the operating room, using tiny instruments to do their work. The surgeon controls the robot from a console, carefully guiding its movements. The robots can perform all sorts of surgeries, from simple procedures like removing a cyst to more complex operations like heart surgery. They can also do things that human surgeons just can't, like getting into tight spaces and seeing inside the body with crystal-clear clarity. The new skills required by surgeons to control robots is another example of why AI won't replace all jobs, but it will empower people to do their jobs better.

The entertainment industry is also set to be transformed by AI, with AI-powered experiences and video games that adapt to our playstyle and preferences. AI will also play a massive role in the world of film and television, with AI-generated movie and TV scripts being accessible. AI can also help with editing and post-production by analyzing footage and suggesting edits, saving time and money. And let's not forget predictive analytics, where AI can analyze audience behavior to predict which movies and TV shows will be successful. This could be a game-changer for studios, allowing them to make more informed decisions about which projects to greenlight.

The general public will begin using completely autonomous vehicles, also referred to as robotaxis, as their primary mode of transportation. Driverless cars have been in development for some time now since the popular Tesla brand emerged as one of the pioneers of this technology. In San Francisco, the public already uses the *cruise* app to call for a driverless car to transport them around. *Waymo* is a similar competitor, and they are like Uber or Lyft but without a human driver!

The next generation of search engines is already on the horizon. *ChatGPT* is a trailblazer in this space, creating more conversational responses rather than a curated list of links that require the user to click on each one to view the results. Google's *Bard* is positioning itself as a competitor to *ChatGPT*. This new iteration of AI search technology is much like having a dialogue. With each new release, these new search platforms will perfect the accuracy of the information and search results. This technology, however, is not 100% accurate and should always be fact-checked for correctness.

More predicitions

The following list from a collection of research studies on Leftronic.com predicts the future of AI:

1. 72% of business leaders see AI as a significant business advantage, and there's more to come.
2. Three-quarters of executives believe AI will help their companies move into new businesses.
3. AI technology could enhance business productivity by up to 40%.
4. AI increases sales numbers by 50% based on the number of leads and this number is expected to grow.
5. 82% of marketers agree that AI and machine learning (ML) will shape the future of marketing.
6. The global AI market is expected to reach $118.6 billion by 2025.

7. The AI software market is expected to hold the largest market share.
8. Investments in AI startups and machine learning are rising in the US.
9. The North American AI chip market is growing at a staggering rate.
10. AI has the productivity and GDP potential of a global economy (M., 2023).

As you can see, the future of AI is bright, exciting, and full of possibilities. And who knows what other incredible advancements it will bring in the years to come since this rapidly growing tech sector has tons of money allocated to it for research and development. The future of AI will get more interesting, with robots that can understand our emotions, homes that run themselves, healthcare that's tailored to our needs, and entertainment that's like nothing we've ever seen. So, hold on tight, and get ready for an AI-enabled ride that's out of this world!

Are you ready? Steps to prepare for future AI advancements and being ahead of the game

AI is no longer just a buzzword in the tech world, but it's becoming a reality in businesses everywhere. It is revolutionizing how companies operate and scale up from small startups to big corporations. But with this exciting new technology comes the challenge of staying on the cutting edge. If you want your company to thrive in the AI revolution, you must be prepared. Let's explore some tips for staying ahead of the AI curve and avoiding obsolescence.

1. *Embrace AI technology.* The first step in preparing for the future of AI is to not be afraid to explore it and form your own opinion. Commit to research before using it. No matter the size of your company, there's an AI solution that can help you improve your operations in some capacity. So, don't fear experimenting with AI technology. There are already tons of AI tools and use cases throughout this book to give you a jumpstart into how you can begin scaling your business with AI today.

2. *Invest in AI talent.* To make the most of AI technology, you need the right people on your team. Investing in AI talent can include hiring AI experts, training your employees, or partnering with AI companies. Whatever you do, make sure you have the right people on board to help you make the most of AI.

3. *Stay informed.* AI is constantly evolving, so staying up-to-date on the latest advancements and best practices is essential. Attend AI conferences, watch videos about it on YouTube, join AI organizations, and read AI publications to stay in the know.

4. *Build a culture of innovation.* To thrive in the AI revolution, you must create an environment where innovation is encouraged. Empower your employees to use AI in new and creative ways, and reward them for their innovative thinking.

5. *Prioritize ethics and transparency.* As AI becomes more widely used, it's necessary to consider the ethical implications of how it's being used. Make sure your company has a solid ethical code in place, and that AI is used according to ethical standards.

6. *Empower your teams.* Don't just introduce AI into your business. Empower your employees to use it effectively. You can

give them the tools and training they need to succeed and encourage them to embrace the technology. Proper training to implement and use AI tools is vital to see your desired results. Feel free to repurpose the talent of your employees to shift their competence towards more AI-driven responsibilities.

7. *Get ahead of the competition.* Use AI to analyze your industry and stay ahead of the competition. Look for ways to use it to differentiate your business and increase efficiency. Become aware of how similar companies to yours are using it effectively.

8. *Foster collaboration.* Finally, foster collaboration between AI experts and your employees. Encourage cross-functional teams to work together to develop new AI solutions and drive innovation. This collaboration is also vital to adopting AI on a company-wide scale because making employees feel a part of the changes is critical. Some employees might resist AI if they see it as a threat to replace their jobs, so be sensitive to this touchy subject.

The most important thing to remember is to start now. The future of AI is moving fast, so don't wait long to begin preparing. I recommend that you start by making a list of problem areas in your business that you want to fix asap. Then, see if any of the tools in this book provides a solution for those problems. Next, start with the most pressing goals you would like to reach and research how these tools can help you reach them. There are so many more tools to explore and many new ones will emerge after this book is published. AI is at your fingertips!

| 12 |

The Time to Supercharge Your Brilliance is Now!

> "In our business, we talk about emerging technologies and how they impact society. We've never seen a technology move as fast as AI has to impact society and technology. This is by far the fastest moving technology that we've ever tracked in terms of its impact and we're just getting started."
> **– Paul Daughterty, Chief Technology and Innovation Officer of Accenture**

The world is rapidly changing right before our eyes, and AI is at the forefront of this transformation. The benefits of it are clear and undeniable. It is already revolutionizing how we live, work, and learn, and we must embrace it now or risk getting left behind. History has shown us that early adopters of new technology are ahead of the competition and give themselves a head start on the way to greater success. Its numerous benefits will help you scale your business in new ways. The urgency of embracing AI cannot be overstated.

Your AI strategy for the future

AI doesn't have to be complex to be effective. A simplified AI strategy can be implemented sooner rather than later. Another key consideration is to start small. It's best to take a strategic and systematic approach, tackling each operational area one at a time. To begin, identify the operational area that would benefit most from an AI solution. This could be anything from streamlining production processes to improving customer service. Once the operational area has been identified, businesses should then gather data related to that area.

Next, it's important to choose the right AI tool or platform that fits the specific needs of the business. There are many off-the-shelf AI tools available that make it easy for businesses to get started with AI without requiring a lot of technical knowledge. Successfully implementing a simplified AI strategy means you'll get to reap the benefits sooner.

Figure 12.1 AI strategy

"Will my business be fine without AI?"

Integrating AI into your operations can seem daunting at first, but remember, this is not the first time we've had to evolve with the changing times. It's unimaginable to think that we actually had a world that existed before smartphones! In the year 2040, we will all look back to today and wonder how we survived without AI! And more importantly, it is crucial to remain innovative and meet the evolving needs of customers. Organizations that do not adopt AI risk being left behind and facing consequences such as losing market share or, worse - becoming obsolete. This is even more critical for women entrepreneurs and business owners because we are already collectively lagging behind our male counterparts, and AI only helps us to compete on a more level playing field.

Loss of competitive advantage. Companies that avoid integrating AI into their operations risk falling behind their competitors that use it to make their operations run more smoothly. As a result, these businesses may need help to keep up with the changing market demands or lose their competitive edge.

Inability to meet customer expectations. Customers today have high expectations for personalized and efficient experiences. Without AI, companies may be challenged to keep up with these expectations, leading to frustrated customers who may take their business elsewhere. Its ease, convenience, and speed give companies an advantage over their counterparts.

Difficulty with data analysis. AI helps organizations collect, manage, and analyze vast amounts of information, allowing them to operate more competently and identify new growth opportunities. Without it, companies may grapple with managing and making sense of their data, potentially missing out on valuable insights and opportunities.

Increased costs and reduced efficiency. The manual procedures completed without AI can be time-consuming and error-prone, leading to increased costs and reduced efficiency. By integrating it into operations, companies can automate predictable job responsibilities, reduce manual effort, and improve accuracy, efficiency, and cost savings.

Stagnant innovation. Without AI, companies may find it difficult to stay ahead of the curve and innovate. AI allows businesses to interpret vast amounts of information and identify new trends and opportunities, enabling them to improve and innovate continuously. Patterns in customer data could reveal new opportunities to expand into new markets, create a new product or service, or shed light on new marketing channels to explore. Finally, integrating AI into your operations is essential to stay on the cutting edge of where your industry is headed. Without it, businesses risk losing competitiveness, meeting customer expectations, difficulties with data analysis, reduced efficiency, experiencing stagnant innovation, and more.

The future is yours!

As technology continues to advance at an unprecedented pace, the benefits of AI are becoming more and more evident. AI can help us solve complex problems, automate routine responsibilities, and provide us with new insights and knowledge. It is literally transforming entire industries, creating new job opportunities, and improving the quality of life for people around the world. However, the benefits of AI are not just some futuristic changes that are coming. It's here now, and the benefits are available to you, but you must seize the opportunity to reap the benefits. Taking advantage of this new way to elevate the way your business functions will position it to scale faster in ways you never imagined!

To embrace AI, we must be proactive in seeking AI-driven solutions and services and explore the many applications of AI. It also means that we must be willing to embrace the change that AI brings and be open to new and innovative ways of doing things. Another reason for the urgency of embracing AI is that it can help us solve some of the world's most pressing problems. AI has the potential to address complex and urgent issues such as environmental issues, poverty, and inequality when used responsibly and ethically.

One of the most significant challenges we face in embracing AI is overcoming our fear of the unknown. Many people are intimidated by the rapid pace of technological changes and the uncertainty that they bring. However, it should be understood that the benefits can far outweigh the risks. The greatest risk is not evolving with the world. Embracing AI also means investing in the technology and the skills and knowledge necessary to use it effectively. It may involve retraining or upskilling existing employees or hiring new employees with the required skills and expertise. However, this investment is necessary to take advantage of the many benefits that it has to offer.

The benefits of AI are too great to ignore, and the future is too important to leave to chance. The AI revolution is likened to the dotcom boom at the turn of the millennium, also referred to as Y2K. It was during this time that many tech innovations emerged, such as Bluetooth, camera phones, and USB flash drives. Those who were early adopters of this technology not only gained a competitive edge in the marketplace, but investors of the tech stocks that exploded made lots of money!

Final thoughts

Congratulations again for investing in yourself and arming yourself with the knowledge and tools you need to supercharge your brilliance and scale your business faster with AI! The information you've just read will help you achieve your goals and reach new heights when applied. By now, I hope you're feeling more confident about AI and all the amazing things it can do for your business. From market expansion to customer acquisition and retention, it can make you take a quantum leap from where your business is now to where you'd like it to be. And the best part? You don't need to be a tech expert to get started. With all the easy-to-use tools and resources available, anyone can use AI today.

As we come to the end of this journey, remember that knowledge only works to your advantage if it's applied. So, take some time to review your company's problem areas and goals and research the AI tools provided to make incremental changes. Don't expect to apply too many tools at once! Roll out a plan on which tools make the most sense for your business to use now and which ones make more sense to use later. Also, remember that AI is a technology sector that grows at lightning speed! Therefore, some tools mentioned will change and have new uses not covered in this book. AI tools will come and go as they continue to evolve, and business owners must keep up with this fast-developing industry to be agile enough to upgrade when necessary.

Throughout this book, we've covered a wide range of topics, from the basics of AI to its real-world applications, to over 100 AI tools for businesses and professionals, to its ethical considerations and best practices. As we look to the future of AI, remember that it is just one of many tools at your disposal.

So, what's next? Well, it's time to put all that knowledge into action! Here are a few steps you can take right now to start using AI in your business:

1. *Get organized.* Make a list of all the areas of your business where you think AI could be helpful. Once you have a list, you can start researching different AI tools and resources to help you achieve your goals.
2. *Start small.* Don't try to tackle everything at once. Choose one area of your business where you want to start using AI, and focus on that. For example, you could start by using it to analyze your customer information and improve your marketing strategy.
3. *Get help.* If you're feeling overwhelmed, don't hesitate to ask for help. Whether you need advice from an AI expert or a mentor in your industry, having someone you can turn to can make all the difference.

Lastly, the key to success with AI is to start small and build from there. Don't be afraid to make mistakes while experimenting. The most important thing is to keep learning and growing, so you can stay ahead of the curve and continue to use what's at your disposal to scale up.

So, there you have it! You now have a more solid foundation to tackle your business goals with AI by your side. So, go forth and knock out those goals one by one! As we bring this part of your learning journey to a close, I wish you all the best in your business endeavors and hope you continue to supercharge your brilliance every day.

Appendix: Decoding the Techie Jargon of AI

Artificial Intelligence (AI) - the simulation of human intelligence in machines designed to think and act like humans. These machines are programmed to perform tasks typically requiring human intelligence, such as recognizing patterns, learning from experience, and making decisions.

Algorithm - a set of instructions for a computer to follow to solve a problem. In AI, algorithms are used to train machine learning models.

AutoML – the automated process of building machine learning models, including data preparation and model selection.

Bias - a systematic error in a model's predictions caused by inaccuracies in the training data or the algorithms used.

Big data - a term used to describe large, complex datasets that traditional data processing methods cannot handle. It refers to the vast amounts of daily information individuals, organizations, and machines generate.

Big data analytics - the use of AI algorithms to sort through and analyze large datasets, allowing organizations to gain insights and make data-driven decisions.

Chatbots - a computer program designed to simulate conversation with human users, especially over the internet. Chatbots use natural language to answer frequently asked questions, schedule appointments, provide customer service, assist with online shopping, and more. They are sometimes referred to as chat agents or virtual assistants.

Cognitive computing - AI focused on building systems that perform tasks requiring human intelligence, such as pattern recognition and decision-making. It involves the creation of systems that can learn, reason, and interact with humans in a way that resembles human intelligence. Cognitive computing systems can perform tasks such as speech recognition, image recognition, and natural language processing.

Computer Vision - a subfield of AI that enables computers to interpret and understand visual information from the world, such as images and videos. It uses algorithms to sort through images and videos, allowing computers to perform tasks including object recognition, image classification, and face recognition.

Data processing - the preparation and sorting of data to use in a machine learning model, including tasks like cleaning, transforming, and normalizing the data.

Deep learning - a subfield of machine learning that uses deep neural networks to model and solves complex problems. It involves using artificial neural networks to model intricate patterns in data. Deep learning algorithms can learn from vast amounts of information and recognize patterns that are too complex for traditional machine learning algorithms.

Explainability - the degree to which humans can understand and interpret AI model predictions. It is like being able to understand why your teacher gave you a particular grade. In other words, it's the ability to understand and explain the decisions made by AI models. Explainability is vital for making sure AI models are accountable and transparent, especially in sensitive matters like medical diagnoses or criminal justice.

Feature engineering - transforming raw data into a format that can be used as input for a machine learning model.

Generative Adversarial Networks (GANs) - a type of neural network used for unsupervised learning, where two neural networks compete with each other to produce realistic output. GANs are neural networks that can create new information that looks like a given dataset. It's like having a copy machine that copies things, but better. It has two parts, a generator, and a discriminator, that work together to make new information look legit.

Generative art - a form of art created using a computer program or algorithm to generate visual or audio output.

Generative pre-trained transformer (GPT) - a type of large language model developed by OpenAI.

Gradient descent - an optimization algorithm used in machine learning to adjust the parameters of a model to minimize the error between the predicted and actual values.

Hyperparameter optimization - the process of selecting the best hyperparameters for a machine learning model, such as the learning rate or the number of hidden layers.

Large Language Models (LLMs) - state-of-the-art artificial intelligence models that have been trained on vast amounts of text data to generate human-like text. These models are like super smart computers that can understand, write, and talk in human language and in several different conversational styles.

Machine learning - a method of teaching computers to learn from data without being explicitly programmed. It is a subfield of AI that uses algorithms to enable machines to learn from information and improve their performance over time. Machine learning algorithms can be trained to recognize patterns in data and make predictions based on them.

Model bias - systematic error in AI models due to imbalanced training data or model design, leading to unfair results. It's like having a favorite in a race. Model bias is when a machine learning model has a mistake because of its training data. It can lead to wrong predictions and discrimination.

Model selection - the process of choosing the best machine learning model for a particular task based on its performance on a validation dataset.

Natural Language Processing (NLP) - a subfield of AI that allows computers to understand, interpret, and generate human language. It is a subfield of AI that deals with the interaction between computers and human languages. NLP algorithms process and understand human language, allowing computers to perform tasks such as text classification, sentiment analysis, and language translation.

Neural network - a computational model inspired by the structure and function of the human brain, made up of interconnected nodes (neurons) that process and transmit information, allowing them to learn from data and make predictions.

Neural Radiance Fields (NeRF) - a type of deep learning model that can be used for various tasks, including image generation, object detection, and segmentation. NeRFs are inspired by the idea of using a neural network to model the radiance of an image, which is a measure of the amount of light that is emitted or reflected by an object.

OpenAI - a research institute focused on developing and promoting artificial intelligence technologies that are safe, transparent, and beneficial to society. One popular AI technology it has created is ChatGPT.

Overfitting - a problem in machine learning where a model is too complex and has learned the training data too well, leading to poor performance on new, unseen data.

Prompt - a piece of text that is used to prime a large language model and guide its generation.

Prompt engineering - prompt engineering is the discipline of developing and refining short text messages, known as prompts, that guide natural language processing (NLP) models to produce desired outputs. The goal is to design the optimal prompt in generative models that produce the most favorable result.

Python - a popular, high-level programming language known for its simplicity, readability, and flexibility. Many AI tools use it.

Reinforcement learning - a type of machine learning where an AI agent learns through trial and error by receiving rewards or penalties for its actions.

Robotics - the branch of technology that deals with the design, construction, operation, and use of robots so that they can perform a wide range of functions that would normally require human intelligence. Robots can be programmed to perform various tasks, from simple monotonous tasks to complex decision-making tasks.

Spatial computing - the use of technology to add digital information and experiences to the physical world. It can include augmented reality, where digital information is added to what you see in the real world, or virtual reality, where you can fully immerse yourself in a digital environment. It has many different uses, such as in education, entertainment, and design, and can change how we interact with the world and each other.

Stable Diffusion - generates complex artistic images based on text prompts. It's an open-source image synthesis AI model available to everyone. Stable Diffusion can be installed locally using code found on GitHub, or users could choose to use other interfaces that already leverage Stable Diffusion models.

Supervised learning - a type of machine learning where the algorithm is trained on a labeled dataset, where the correct output is already known. It's like exploring a new city with a map. It's used to make predictions based on the relationship between inputs and outputs in the training data, and it's often used in classification and prediction.

Transfer learning - a technique in deep learning where a pre-trained model is fine-tuned for a different but related task. It's like using your knowledge from one class in another class. This way, the model can use what it previously learned to do better in the new task.

Unsupervised learning - a type of machine learning where the algorithm is given a dataset without any labels and must find patterns or relationships on its own. It is similar to exploring a new city without a map. It's used to find patterns and structure in data, and it's often used in things like clustering and reducing data.

Webhook - A way for one computer program to send a message or information to another program over the internet in real time. It sends the message to a specific URL that belongs to the other program. Webhooks are often used to streamline processes and make it easier for different programs to communicate and work together. They are helpful for developers who want to build custom applications or create integrations between other software systems.

References

Amazon. (2018, July 9). Protecting data privacy. Retrieved from https://www.aboutamazon.com/news/amazon-ai/protecting-data-privacy

Crampton, N. (2021, January 19). The building blocks of Microsoft's responsible AI program. Microsoft On the Issues. Retrieved from https://blogs.microsoft.com/on-the-issues/2021/01/19/microsoft-responsible-ai-program/

Cutler, A., & Pribić, M. (2022, December 6). Everyday ethics for AI. IBM Design. https://www.ibm.com/design/ai/ethics/everyday-ethics/.

Daugherty, P. (2021, September 17). Accenture Comments to National Institute of Standards and Technology Artificial Intelligence Risk Management Framework. Retrieved from https://www.nist.gov/system/files/documents/2021/09/17/ai-rmf-rfi-0098.pdf

Goergen, A. (2022, November 16). Artificial Intelligence & Intelligent Automation in Sales: Use cases and Best Practices. Levity. https://levity.ai/blog/artificial-intelligence-and-intelligent-automation-in-sales.

Google. (2023, February 11). Artificial Intelligence at Google: Our Principles. Retrieved from https://ai.google/principles/.

M., M. (2023, March 07). 27 Artificial Intelligence Statistics: What's New in 2023? Leftronic. https://leftronic.com/blog/artificial-intelligence-statistics/.

National Center for Women & Information Technology (NCWIT). (2019). Women in Small Business: The Value of Technology Adoption. https://ncwit.org/resources/research-stats/.

National Women's Business Council. (2021). The Impact of COVID-19 on Women's Small Businesses. NWBC. https://www.nwbc.gov/.

Rubin, C., Hakspiel, J., & Gray, B. (2021, February 23). Using Artificial Intelligence and Technology for Women's Economic Empowerment: How Can It Work? SEEP Network. https://seepnetwork.org/Blog-Post/Using-Artificial-Intelligence-and-Technology-for-Women-s-Economic-Empowerment-Can-It-Work.

Sage. (2019). Sage SMB Survey on AI. Sage. https://www.sage.com/.

World Economic Forum. (2022, August 22). Why We Must Act Now to Close the Gender Gap in AI. https://www.weforum.org/agenda/2022/08/why-we-must-act-now-to-close-the-gender-gap-in-ai/.

Zipkin, N. (2015, April 17). 8 Far-Out 'Jetsons' Contraptions That Actually Exist Today. Entrepreneur. Retrieved from https://www.entrepreneur.com/growing-a-business/8-far-out-jetsons-contraptions-that-actually-exist-today/245192.

Tamiko Cuellar is a dynamic and inspiring entrepreneur and International Business Strategist with a proven track record of global impact. She has completed diplomatic missions in partnership with the U.S. Embassies in South Africa and Namibia on business development, reducing unemployment and poverty, and international trade. She was invited to the State House in Namibia to consult on its country's entrepreneurship initiatives. Tamiko was also selected as a Fulbright Specialist by the U.S. Department of State to develop entrepreneurs in Namibia and helped to launch the first entrepreneurship incubator at the University of Namibia's Business School (UNAM).

Tamiko has trained emerging entrepreneurs at the Kwame Nkrumah University of Science & Technology (KNUST) in Ghana and the University of South Africa (UNISA) and served as a guest lecturer for MBA students at UNAM. She earned three college degrees in business and served as a contributing writer for Forbes.com® and The Huffington Post. Additionally, she has been an expert media guest on major media outlets in several African countries and is an international keynote speaker.

Tamiko's rise to success is truly inspiring. Born into a broken family in low-income housing projects, she grew up in poverty in a crime-ridden neighborhood where she attended underperforming schools in East St. Louis, Illinois. Despite these challenges, Tamiko had a successful and adventurous career in corporate America before launching her own companies, where she helped to secure over $30 million in investment revenue.

Tamiko is the CEO and founder of Pursue Your Purpose LLC, a global firm that coaches aspiring and emerging women entrepreneurs and corporate intrapreneurs. She continues to use her unique approach and methodologies to align purpose with profits for thousands of entrepreneurs and business leaders around the globe. Her company provides live and virtual courses, coaching strategy sessions, and resource books. Tamiko is also the author of *Cultivating an Entrepreneurial Mindset* and *Own Your Brilliance! — A Woman's Guide to Hiring Herself.*

Tamiko resides in St. Louis, MO, where she continues making her mark on both the local and global business communities. Her expertise, experience, and distinctive style of entrepreneurship have helped thousands of women and emerging business leaders worldwide profit from their purpose and scale their businesses. Her rise from struggle to becoming an impactful entrepreneur is remarkable and demonstrates a powerful example to women business owners, entrepreneurs, corporate leaders, and women of color across the globe. For media appearance requests, booking for speaking engagements, or group training inquiries, email info@PursueYourPurpose.com.

PHOTO GALLERY

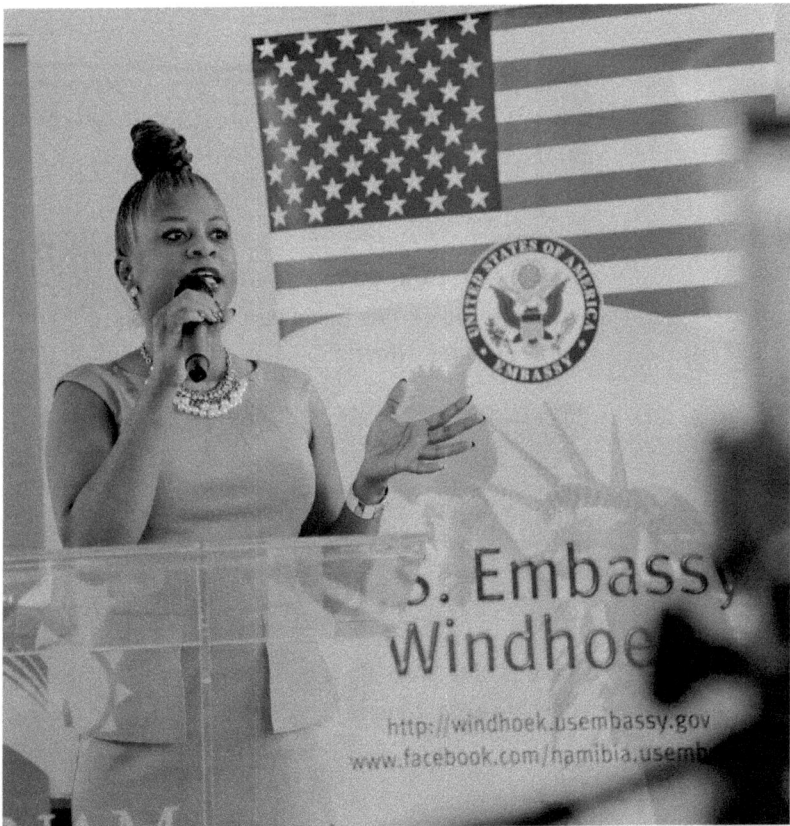

Keynote speaker for the U.S. Embassy in Namibia on women in leadership

Presenting my book "Own Your Brilliance! A Woman's Guide to Hiring Herself" to the U.S. Embassy in Ghana

Guest on "Ladies Talk Business Show" in Lagos, Nigeria

Guest on the "Wake Up Nigeria" show in Lagos, Nigeria

Speaking with the former U.S. Ambassador to Namibia, Lisa Johnson

Inspiring entrepreneurs during a diplomatic mission to South Africa

4th published book, "Cultivating an Entrepreneurial Mindset," 2019

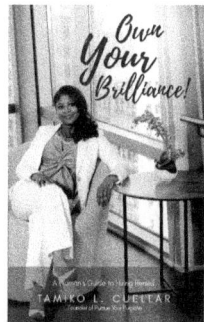

3rd published book, "Own Your Brilliance!," 2017

www.ingramcontent.com/pod-product-compliance
Lightning Source LLC
Chambersburg PA
CBHW071703210326
41597CB00017B/2311